T0360886

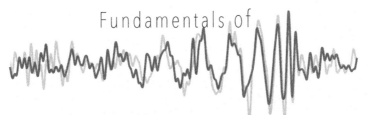

Fundamentals of
Interferometric
Gravitational Wave
Detectors

Second Edition

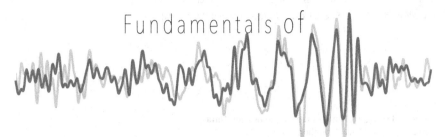

Interferometric
Gravitational Wave
Detectors

Second Edition

Peter R. Saulson

Martin A. Pomerantz '37 Professor of Physics
Syracuse University, USA

World Scientific

NEW JERSEY · LONDON · SINGAPORE · BEIJING · SHANGHAI · HONG KONG · TAIPEI · CHENNAI · TOKYO

Published by

World Scientific Publishing Co. Pte. Ltd.

5 Toh Tuck Link, Singapore 596224

USA office: 27 Warren Street, Suite 401-402, Hackensack, NJ 07601

UK office: 57 Shelton Street, Covent Garden, London WC2H 9HE

Library of Congress Cataloging-in-Publication Data
Names: Saulson, Peter R., author.
Title: Fundamentals of interferometric gravitational wave detectors / Peter R. Saulson
    (Syracuse University, USA).
Description: 2nd edition. | Hackensack, NJ : World Scientific, [2017] |
    Includes bibliographical references and indexes.
Identifiers: LCCN 2016041682| ISBN 9789813143074 (hard cover ; alk. paper) |
    ISBN 981314307X (hard cover ; alk. paper)
Subjects: LCSH: Gravitational waves. | Interferometers.
Classification: LCC QC179 .S28 2017 | DDC 539.7/54--dc23
LC record available at https://lccn.loc.gov/2016041682

British Library Cataloguing-in-Publication Data
A catalogue record for this book is available from the British Library.

Desk Editor: Ng Kah Fee

Typeset by Stallion Press
Email: enquiries@stallionpress.com

Printed in Singapore

Contents

List of Figures

Preface to the Second Edition

The discovery of gravitational waves by the two LIGO interferometers on 14 September 2015 sent me back to this introductory book on how to understand interferometric detectors of gravitational waves. It seemed likely that there might be a new generation of students interested in the book.

A complete revision was not possible. Instead, I've written an epilogue (Chapter 17), intended to show how to use the original book as an introduction to the subject, and then how a reader can move on to a higher level of competence by study of the more recent (and more advanced) literature. I urge the reader to consult Chapter 17 early and often. That chapter's introduction (Section 17.1) should be read right away for further guidance on how this edition should be used today; then, as the reader progresses in her study of the book, I'd urge her to return to the relevant sections of the epilogue for commentary, context, and suggestions for further study.

I've spent the entire period since writing the first edition as a member of the Department of Physics in the College of Arts and Sciences at Syracuse University. It has remained the most congenial and supportive home that one could ask for. Members of the SU Gravitational Wave Group have made it a pleasure to come to work every day. I am profoundly grateful to my former graduate students Gabriela González, Yinglei Huang, Will Startin, Andri Gretarsson, Scott Kittelberger, Eiichi Hirose, Matt West, and T. J. Massinger. Among the postdocs with whom I've worked, I'd like to thank especially Mark Beilby, Gregg Harry, Steve Penn, Alessandra Di Credico, Josh Smith, Antonio Perreca, and Laura Nuttall, as well as my colleague Ryan Fisher. My fellow faculty members, Duncan Brown and Stefan Ballmer, have brought life, leadership, and vision to the group. All of the members of our group have been a tremendous source of understanding and insight of all kinds.

Throughout that time, I have been privileged to receive generous support from the National Science Foundation, most recently under grant PHY-1607169. The NSF's faithfulness to our long struggle to detect gravitational waves deserves the most profound thanks.

Over these years, I've been privileged to work in community with the most wonderful set of colleagues one could possibly hope for. The LIGO Scientific Collaboration (including the outstanding team of the LIGO Laboratory) is a valiant band that has journeyed together on a long joint quest. It has been a joy to work with you all.

The most joyful constant of these years has been the love and support of my wife Sarah F. Saulson; words cannot express my debt to her.

One sad change since 1994 has been the passing of my parents, Dr. Stanley H. Saulson and Helen M. Saulson, to whom I owe so much. This book is dedicated to their memory.

Syracuse University
July 2016

Preface

This book is an attempt to communicate the basic logic of interferometric gravitational wave detectors to people who are new to the field. With the recent start of construction of the Laser Interferometer Gravitational Wave Observatory in the United States, and similar ventures elsewhere, it seems likely that many people will want to know more about the subject, including both young people at the beginnings of their scientific careers and experienced scientists with a new interest in an expanding field. While I in no way expect that this book will displace the more specialized literature, I hope it will serve as an overview of a subject of tremendous intellectual excitement and technical sweetness. I assume the reader has a basic knowledge of physics, but no special familiarity with either general relativity or the special techniques of experimental physics.

> ... a grateful mind
> By owing owes not, but still pays, at once
> Indebted and discharged; what burden then?
>
> Paradise Lost, IV:55-7

The inspiration for this book came from Abhay Ashtekar. As he promised, the Physics Department at Syracuse University has proven to be a wonderful home at which to teach, to learn, and to write. The outline of this book was worked out during a seminar I conducted there with my research group. Large parts of the manuscript were read carefully by Stan Whitcomb and by Gabriela Gonzalez; crucial parts were also read by John Price, David Shoemaker, and Robert Spero. Yinglei Huang, Alex Evako, Joe Adams, and especially Gabriela Gonzalez were of tremendous help in improving the presentation, tracking down important facts and ferreting out mistakes. Bess Collins taught me how to look at the figures from the point of view of a graphic designer. Everyone named above made valuable suggestions for which I am grateful. I am of course responsible for all of the mistakes that remain.

I was privileged to be able to present the contents of Chapters 2 and 5 of the present volume as lectures in the School of Cosmology and Gravitation at Erice in May, 1993. My thanks go to Joseph Weber for the invitation to lecture there, and to the School's organizer Venzo de Sabbata and the staff at Erice for creating such a pleasant ambiance for the study of physics.

It is with profound gratitude that I thank here those who have helped to shape me as a physicist. I was inspired to study physics by Edward Purcell, and guided along the way by Paul Bamberg, David Layzer, and Herbert Gursky. My graduate apprenticeship was in the hands of David Wilkinson, a model of what a scientist should be. In the Gravity Group at Princeton, the inspiring example of Robert Dicke was never far away. I learned the trade of gravitational wave detection from Rainer Weiss; Rai not only initiated me into the mysteries of the field, but also set standards of intellectual rigor and personal responsibility that are a continuing model.

I have also learned a great deal from the graduate students alongside of whom I have been privileged to work; at MIT, this included Dan Dewey, Jeff Livas, David Shoemaker, Michelle Stephens, Joe Kovalik, Nelson Christensen, Peter Fritschel, and Joe Giaime. Throughout my career, it has been my privilege to be surrounded by colleagues and students who were a constant source of insight and inspiration; the talent and openness of the members of the gravitational physics community are exemplary. Thanks to you all.

This work was supported in part by the National Science Foundation, through grant PHY-9113902. I am grateful to the American taxpayers for underwriting this field of research, and for the responsible care exercised on their behalf by Richard Isaacson.

For support in all possible forms over many years, I thank my parents, Dr. Stanley H. Saulson and Helen M. Saulson. I owe my deepest thanks to my wife, Sarah F. Saulson, for her love, encouragement, and fine editorial ear.

The Search for Gravitational Waves　　1

The fact is the sweetest dream that labor knows.

Robert Frost, "Mowing"

The next few years should see the commissioning of several new gravitational wave detectors of unprecedented sensitivity. There is real excitement for the Laser Interferometer Gravitational-Wave Observatory (LIGO) in the U.S., Virgo and GEO in Europe, and all of the other gravitational observatories that may soon be built. While ultimate success is not guaranteed, the prospects are good that the world will soon be equipped with a network of gravitational wave detectors sensitive enough to record numerous signals of astronomical origin, broadband enough to allow waveform analysis that may reveal the structure of the sources, and widespread and redundant enough to allow location of the sources on the sky by triangulation. With such a functioning network, it should be possible to experimentally verify the basic physics of gravitational waves, as predicted by the General Theory of Relativity. Furthermore, it is quite likely that we will be able to open a new "gravitational wave window" in astronomy, revealing facets of the Universe previously unseen with electromagnetic waves.

The quest for gravitational waves is a true 20th century scientific adventure.

This book is intended to serve as an introduction to that quest. It is written by an experimental physicist, for those interested in learning more about the quest as an adventure in experimental physics. Before we set out, it is worth a brief look at where the search for gravitational waves fits into the history of gravitational physics, and where it fits into the practice of contemporary physics and astronomy.

1

1.1 The Importance of the Search

Gravitation has always been a subject in which new experimental data are hard to come by. This makes any new way of learning about gravity a potentially precious resource. In the introduction to his 1964 book, *The Theoretical Significance of Experimental Relativity*, Robert H. Dicke commented on this situation[1]:

> The problems of the experimentalist working on gravitation differ from [those in other fields]. Here there is an elegant, well defined theory but almost no experiments. The situation is almost orthogonal to that in both high energy physics and solid state physics. Far from experimental science being a crutch on which theory leans, in the case of general relativity, theory has far outstripped experiment, and the big problem is one of finding significant experiments to perform. This situation raises serious problems for theory and our understanding of gravitation. For where there are no experiments the theory easily degenerates into purely formal mathematics.
>
> ...In other fields of physics the experimentalist is faced with the problem of choosing the most important out of a large number of possible experiments. With gravitation the problem is different. There are so few possible experiments, and their importance is such, that any and all significant experiments should be performed.

Dicke next describes the role of *null experiments* (see our Chapter 10) and the classic laboratory and solar system tests of relativity (see our discussion in the next section), then continues:

> ... The third [kind of experiment], which could be easily rocketed into first place of importance by a significant discovery, is the general area of cosmological effects. These concern a wide variety of subjects: possible gravitational or scalar waves from space, the effects of a quasi-static scalar field generated by distant matter (if such a field exists), the continuous creation cosmology (which should be either laid to rest or else cured of its ills), Schwarzschild stars if they exist (a Schwarzschild star is one for which the star's matter is falling down the gullet of the Schwarzschild solution), massive generators of gravitational radiation, global aspects of the structure of the Universe, and the origin of matter. All these and many more are subjects concerning effects on a cosmological scale. These effects, if they could be observed, would give information about the most fundamental of the problems of physics.

At least three of the six entries in Dicke's list of potentially "first place" discoveries involve either directly or indirectly (in the case of "Schwarzschild stars", now referred to as *black holes*) the search for gravitational waves. We will discuss below how it comes to be that gravitational waves play such a crucial role in our search for further empirical understanding of gravitation.

1.2   A Bit of History

The concept of a gravitational wave comes from the work of the most revered physicist of modern times, Albert Einstein. After revamping mechanics and electrodynamics with his Special Theory of Relativity, Einstein saw that Newtonian gravitation theory was also in need of revision. In relativistic physics, the essence of the notion of *causality* is the prohibition against sending any signal faster than the speed of light. Yet the instantaneous action-at-a-distance of Newton's picture of gravity would let an advanced civilization operate a gravitational radio system with no propagation delays at all. When the General Theory of Relativity, his beautiful theory of gravity, showed instead that fluctuating gravitational fields would lead to waves that could only just keep pace with their electromagnetic cousins, Einstein must have been sure his ideas had gone in the right direction.

The dual role of mass, as both measure of inertia and as gravitational charge, had puzzled Newton and his successors. General relativity's deepest motivation was to show how this could come about naturally and exactly. The crucial insight was that what we normally call the "force of gravity" isn't a force separate from the space-time arena in which it acts; instead, it is a reflection of the fact that objects move along the geodesics of curved space-time. The gravitational field of a massive body consists of the way in which it curves space-time. Bodies travel through space-time along paths that are as straight as possible, given the curvature; that those paths aren't truly straight represents what we would conventionally describe as the gravitational acceleration.

This picture of gravity as space-time curvature embodies general relativity's profoundest insight. But general relativity also made predictions of specific observable phenomena. It resolved the long-standing puzzle of the precession of the planet Mercury's not-quite elliptical orbit. Eddington's verification of the prediction that the Sun's gravitation could bend even starlight made Einstein a celebrity around the world in 1919.[2] Another "classic test" of relativity, that clocks deep in a gravitational potential well should appear to run slowly as seen by an observer outside, wasn't verified until the 1960s.

Compared to these experimental triumphs, the concept of gravitational waves remained an oddity — a crucial idea of great importance for the integrity of the theoretical structure, yet seemingly so weak in its visible manifestations as to remain forever beyond direct observation and test. Einstein himself had no confidence that gravitational waves would ever be detected. For a time some theorists even thought the wave solutions to Einstein's field equations might be made to vanish by suitable coordinate transformations, in which case gravitational waves as physical objects would not exist at all.

Around 1960, Joseph Weber[3] began the struggle to make gravitational waves amenable to experimental study. He set out to monitor the minute vibrations of massive aluminum cylinders, hoping to see them set into oscillation by passing gravitational waves. It was an effort that must surely have seemed to his colleagues at the time either remarkably courageous or exceptionally foolhardy. All the more reason that Weber's 1969[4] announcement that his pair of detectors were registering coincident signals, likely of astronomical origin, caused waves of excitement to spread rapidly through the scientific community.

A discovery of gravitational waves demanded independent verification. In the effort, the community of gravitational wave experimenters was born. Sadly, what they found was that no one else's instruments registered the signals that Weber saw. Weber himself has never accepted that verdict, but by the mid '70s the consensus was that whatever gravitational waves existed were in fact too feeble for the sensitivity of that era's instruments. For a review of the history of this first phase of the search for gravitational waves, see the article by Tyson and Giffard,[5] and our own discussion, including the most recent upper limits on gravitational wave strengths, in Chapter 15.

The dashing of the hopes that Weber's announcement had encouraged led some to abandon the quest. Others redoubled their efforts, but the work has not been easy. It is only now, in the mid 1990s, that technology has improved to the point that the "smart money" is riding on predictions of successful detections of gravitational waves in the near future.

What is the reason for our present optimism? It is one of the goals of this book to provide an answer to that question. There has been much progress in building improved successors to Weber's original *resonant mass detectors*, colloquially known as *bars*. (For a discussion of this technology, and a brief survey of the present state of the art, see Chapter 13.) Even so, our greatest hopes are now fixed on a different sort of detector, involving large *interferometers*. These are instruments much like that used by Michelson and Morley to look for the *ether drift*. The most obvious differences are that the mirrors that define the light paths will be several kilometers apart, instead of a few meters, and will hang freely on pendulums instead of being fixed to a rigid structure.

In the rest of this book, we will discuss the physical principles of the detection of gravitational waves. We will compare the sensitivity of instruments we plan to build with the strengths of waves the Universe is likely to provide us. Those plans are embodied in the LIGO Project (for Laser Interferometer Gravitational-Wave Observatory) in the United States,[6] the Italian-French Virgo Project,[7] and similar initiatives of scientists around the world. Construction of these large interferometric detectors is just now getting under way. At appropriate points throughout this book, we will examine what features these instruments will need to have in order to successfully detect gravitational waves. The book will conclude with a brief overview of the plans for LIGO, with the aim of showing why there is optimism that fulfillment of this quest is near at hand.

1.3 The Practice of Gravitational Wave Detection

Those who work in the field of gravitational wave detection are heirs to several great traditions. One is the line of high precision mechanical experiments, exemplified by the work of Cavendish, Eotvos, and Dicke. At the heart of any gravitational wave detector are mechanical test bodies that must be painstakingly constructed and carefully shielded from external disturbances, so that they may sufficiently resemble an ideal test mass. Failure in this respect would make the results of the experiment hard to interpret or, much worse, hopelessly dominated by uninteresting noise effects.

Another line of forebears also paved the way for the eventual success of gravitational wave detection. The techniques of precision optical measurements, starting with Michelson, and strongly augmented by the developers of the laser, are crucial to the attainment of unprecedented sensitivity to very small mechanical effects. The communications engineers of the pre-War period, and the microwave pioneers of World War II (including, here as well, Dicke), contributed many important insights and tricks to the operation of ultra-precise measurement systems.

The participants in the enterprise of gravitational wave detection must be masters, collectively if not always individually, of this vast body of expertise in experimental physics. But in addition, the work must be continually guided by an understanding of modern astronomy and relativistic astrophysics. For reasons we will discuss later, no terrestrial generator of gravitational waves seems practicable. To find gravitational waves, we are forced to look to the stars.

This feature of the field places it squarely in the mainstream of post-War astronomy. That science used to rely totally on visible light as a carrier of information from distant objects. Since the 1940s, other bands of electromagnetic radiation have been pressed into service. Now the useful forms of electromagnetic waves span the entire spectrum from radio waves, through infrared and ultraviolet light, to X-rays and

γ-rays. In many instances, new ways of looking at the sky have led to the discovery of phenomena of radically distinct and previously unsuspected kinds. Thus, the technological revolution in measurement has produced what is in some ways a Golden Age of astronomy. New discoveries continue to occur at an exciting pace, as continued technical development leads to vastly improved performance of sensors of many kinds. The trend has been so successful that it has almost come to be taken for granted.

An interesting analysis of this historical process was presented by Martin Harwit in his book *Cosmic Discovery*.[8] He showed how large a fraction of important new results throughout the history of astronomy could be directly traced to innovations in instrumentation that gave qualitatively new or substantially improved sensitivity to cosmic signals. He went on to prescribe a way to plan for maximizing the rate of future astronomical discoveries, by concentrating on the development of substantially improved or qualitatively new methods of looking at the sky. Not surprisingly, the development of sensitive detectors of gravitational waves, an information carrier completely distinct from electromagnetic radiation, was included in Harwit's short list of projects with outstanding prospects for success.

Harwit also considered the question of what sort of people made the dramatic new discoveries in astronomy. His study showed that, in most cases, the rewards for developing new technology went directly to those who themselves had built the observing instruments. Partly this must of course be due to having first access to new techniques. But, he claimed, it also reflects the fact that those who have the deepest competence in a new technique are the ones who can have the greatest confidence in its results. They are the people who can spot and debug problems as they occur, and who can therefore also firmly establish the validity of surprising results.

To the extent that one can profitably draw lessons from history, this argument is a powerful encouragement to persevere in the long effort to find gravitational waves. Now that success appears likely to come soon, the attractions of this field must seem even stronger.

1.4   A Guide for the Reader

In the rest of this book, I will attempt to explain, as clearly as possible, the logic of the efforts to detect gravitational waves. We will study the fundamental quantum mechanical and thermodynamic limits to the precision of the measurements. I will also give an introduction to some of the truly "sweet" technology that lies at the heart of our present efforts to detect gravitational waves. And I will provide an overview of some of the exciting aspects of relativity and astrophysics, including black holes, neutron stars, and the Big Bang itself, that form both the context and the ultimate subject matter of this field.

The belief governing my choice of topics is that the successful detection of gravitational waves will only be achieved if we apply knowledge from across the spectrum of experimental physics. For this reason, I have devoted space to a discussion of some fundamentals with much wider applicability than to this experiment alone. Substantial sections of the book are devoted to the basics of linear system theory, power spectral methods, feedback control, null experiment design, and extraction of signals from noise. Readers who already have knowledge of these topics will be able to skip those parts, although briefly scanning them to become familiar with my choice of notation might be worthwhile. For newcomers to experimental physics, these sections cannot of course replace specialized texts. But I hope they can give an overview that is helpful in following the discussion of gravitational wave detection. I also hope that these sections can help students to learn how to apply these basic techniques after they have studied them in textbooks, by supplying a set of examples beyond those in the standard literature, focused on the needs of this particular experiment. The material in Chapter 10 on null instrument design may prove especially useful in this regard; although it is part of the standard wisdom of experimental physics, I know of no reference where the ideas have been set down in such a systematic way.

This book's goal is to communicate the basic ideas of the field at a level appropriate to serious newcomers. I have learned a great deal in the process of writing the book, and so I hope that even experts might find a few ideas they haven't come across before, or at least see something said in a fresh way. But a book such as this can only scratch the surface of the field. For more detailed information, the reader must turn to the more specialized literature, or to other recent reviews with different emphases. Among the best of the latter for the theoretical issues is Kip Thorne's chapter called "Gravitational Radiation" in *300 Years of Gravitation*, edited by Hawking and Israel.[9] An expanded version is soon to appear as a separate volume. For experimental matters at a more detailed level than is presented here, David Blair's excellent *The Detection of Gravitational Waves* is recommended.[10] Both books also contain extensive lists of references.

In brief, the plan of the book is as follows. Chapters 2 and 3 introduce the concept of gravitational waves, explain at the level of a *gedanken* experiment how they might be detected, and survey how they are generated. Chapter 4, the first of the chapters devoted to general techniques of experimental physics, discusses the basics of time series analysis, linear system theory, and the concept of the signal-to-noise ratio. Chapters 5 and 6 examine the fundamental limits to the sensitivity of interferometric measurements, while Chapters 7 and 8 are addressed to the most important sources of mechanical noise. Chapter 9 discusses a couple of important features of multi-kilometer interferometers. Chapters 10 and 11 survey the topics of null instrument design and of the operation of feedback systems, crucial both for gravitational wave interferometers and for many other physics experiments. Chapter 12 applies this material to a more detailed description of interferometer operation. The principles of resonant mass detectors of gravitational waves, the chief alternative to interferometric detectors, are surveyed in Chapter 13. The fundamentals of extraction of gravitational wave signals from noise are discussed in Chapter 14. Astronomical interpretation of the signals is the topic of Chapter 15, where we also summarize the results of experiments carried out to date. Finally, Chapter 16 shows how new instruments soon to be constructed, as exemplified by the LIGO project, are expected to fulfill the promise of the field.

The book pays more attention than is customary in introductory texts to the historical antecedents of present-day experiments. It is my belief that we can draw both technical and moral inspiration from the successful struggles of our predecessors. The question of sorting out the recent history of the development of gravitational wave interferometers is more problematic. This is a field in which free discussion and informal dissemination of ideas have been better developed than formal publication, making the origin of many specific ideas difficult to ascertain. I have tried to attribute individual contributions to the best of my ability.

The Nature of Gravitational Waves 2

The existence of wave solutions to Einstein's field equations is, arguably, the most "relativistic" feature of the theory of gravitation known as the General Theory of Relativity. This is true in the sense that it is the instantaneous action-at-a-distance of Newtonian gravity that most offends our notions of causality as understood in the context of the Special Theory of Relativity. Newton's theory, taken literally, would predict that the gravitational field produced by a mass always has the familiar $1/r^2$ form, with r referring to its present position, no matter how rapidly it might move (or accelerate), and no matter how far away from it we consider its field.

General relativity fixes the problem posed by moving sources of a gravitational field. It proposes that the gravitational field (that is, the curvature of space-time) does not change instantaneously at arbitrary distances from a moving source. Instead, in a manner deeply analogous to electromagnetic waves, the "news" of the motion of a source of space-time curvature propagates at the speed of light.

In this chapter, we will briefly examine how gravitational waves of finite speed arise in general relativity. Then we will discuss the interaction of gravitational waves with systems of test bodies, and consider how such interactions may be observed.

2.1 Waves in General Relativity

One of the most fundamental concepts in the Special Theory of Relativity is that the *space-time interval ds* between any two neighboring points is given by the expression

$$ds^2 = -c^2 dt^2 + dx^2 + dy^2 + dz^2 \tag{2.1}$$

or

$$ds^2 = \eta_{\mu\nu} dx^\mu dx^\nu, \tag{2.2}$$

with the Minkowski metric $\eta_{\mu\nu}$ given, in Cartesian coordinates, by

$$\eta_{\mu\nu} = \begin{pmatrix} -1 & 0 & 0 & 0 \\ 0 & 1 & 0 & 0 \\ 0 & 0 & 1 & 0 \\ 0 & 0 & 0 & 1 \end{pmatrix}. \tag{2.3}$$

Note that in Eq. 2.1 all of the superscripts indicate raising to the second power, while in Eq. 2.2 the Greek indices range from 0 to 3 to represent t, x, y, and z, respectively. Equation 2.2 also makes use of the famous "repeated index summation convention".

The same physical concept is carried over into the General Theory of Relativity, with one key difference — space-time is no longer necessarily the "flat" space-time described by the Minkowski metric, but will in general be curved in order to represent what we call gravitation. The more general statement of the definition of the space–time interval is

$$ds^2 = g_{\mu\nu}dx^\mu dx^\nu, \tag{2.4}$$

where all of the information about space-time curvature is encoded in the metric $g_{\mu\nu}$. A lot of physics can be expressed in Eq. 2.4. Fortunately, we only need to work in one simple special case, that of a small perturbation to flat space-time. Then it makes sense to write the metric in the form

$$g_{\mu\nu} = \eta_{\mu\nu} + h_{\mu\nu}, \tag{2.5}$$

where $h_{\mu\nu}$ represents the *metric perturbation* away from Minkowski space.

The key physics of the problem is thus carried in the form of $h_{\mu\nu}$. It is a remarkable fact that in the weak-field limit, the non-linear Einstein equations can be approximated as linear equations. There is a great deal of *gauge freedom* in the construction of explicit forms for $h_{\mu\nu}$, and this is the source of some confusion. But there is one particularly useful gauge choice in which both physics and mathematics become especially clear. This is the so-called *transverse traceless gauge*, or "TT gauge" for short. In this gauge, coordinates are marked out by the world lines of freely-falling test masses. With this choice of coordinates, the weak field limit of Einstein's field equation becomes a wave equation[11]

$$\left(\nabla^2 - \frac{1}{c^2}\frac{\partial^2}{\partial t^2} \right) h_{\mu\nu} = 0. \tag{2.6}$$

Thus, the elements of $h_{\mu\nu}$ can take the form $h(2\pi f t - \mathbf{k} \cdot \mathbf{x})$, with $f = |\mathbf{k}|/2\pi c$, representing a plane wave propagating in the direction $\hat{k} \equiv \mathbf{k}/|\mathbf{k}|$ with the speed c.

The appearance of the speed of light as the speed of gravitational waves is a result of the way space and time are brought together in relativity as space-time. As waves of space-time disturbances, c is the only "natural" speed in the problem.

Consider the case of a wave propagating along the $\hat{z}$ axis. The statement that $h_{\mu\nu}$ be transverse and traceless means that it has the form

$$
h_{\mu\nu} = \begin{pmatrix} 0 & 0 & 0 & 0 \\ 0 & a & b & 0 \\ 0 & b & -a & 0 \\ 0 & 0 & 0 & 0 \end{pmatrix}.
\tag{2.7}
$$

In other words, we can write this wave as a sum of two components, $h = a\hat{h}_+ + b\hat{h}_\times$, with

$$
\hat{h}_+ = \begin{pmatrix} 0 & 0 & 0 & 0 \\ 0 & 1 & 0 & 0 \\ 0 & 0 & -1 & 0 \\ 0 & 0 & 0 & 0 \end{pmatrix},
\tag{2.8}
$$

and

$$
\hat{h}_\times = \begin{pmatrix} 0 & 0 & 0 & 0 \\ 0 & 0 & 1 & 0 \\ 0 & 1 & 0 & 0 \\ 0 & 0 & 0 & 0 \end{pmatrix}.
\tag{2.9}
$$

The "basis tensors" $\hat{h}_+$ and $\hat{h}_\times$ (pronounced "h plus" and "h cross") represent the two orthogonal polarizations for waves propagating along the $\hat{z}$ axis. This is true in spite of the fact that one goes into the other through a rotation of only 45°. We can make this intuitively clear by drawing a diagram, as shown in Figure 2.1, in which we graph how the distances between a set of points originally arranged in a pair of squares are perturbed by $h_{\mu\nu}$. The perturbation called h_+ momentarily lengthens distances along the $\hat{x}$ axis, simultaneously shrinking them along the $\hat{y}$ axis. The polarization $h_\times$ has its principal axes rotated 45°. Clearly, a 90° rotation takes one of these tensors into itself, up to a sign change. A particle physicist would say this is a reflection of the fact that the *graviton* is a spin-2 particle. We will have nothing else to say about gravitons — a classical picture is perfectly adequate for the situations we will be describing. The most general form for $h_{\mu\nu}$ can be generated by considering arbitrary spatial rotations of h_+ or $h_\times$.

It is important to understand how gravitational effects appear when we describe events in such a coordinate system. The TT gauge is a coordinate choice that embodies the deepest principle of general relativity: gravitation is not a force, but

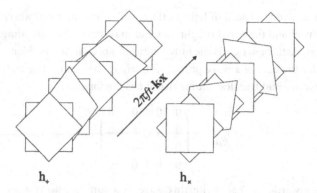

$2\pi f t - k \cdot x$

h_+ $h_\times$

Figure 2.1: An illustration of the two polarizations of a gravitational wave. The effect of a wave is shown by how it changes the distances between a set of freely-falling test masses arranged in two squares.

a phenomenon of geodesic motion through curved space-time. This means that, from the relativistic point of view, a freely-falling mass (by definition an object that is subject to no influences of non-gravitational origin) does not accelerate when a gravitational wave passes. It just sits there. So, we can use a set of freely-falling masses to define a coordinate system in space-time.

This is a beautiful idea, but it immediately leads to two related questions. The first is, "If test masses don't respond to a gravitational wave, how can there be any observable physical manifestation of a gravitational wave?" This is tantamount to asking whether gravitational waves are "real" in any meaningful sense. We will find the answer to this question by examining how changes in the metric can affect the measurable distance between our freely-falling test masses. The next two sections will explore this answer.

The second question has to do with making the translation between this relativistic language and the more conventional description of gravitation in the laboratory and in everyday experience. (Try going to the weight room at the gym and telling people there that gravity exerts no forces!) The final section of this chapter will discuss under what circumstances it can make sense to use the Newtonian concept of a gravitational force to describe the effects of a gravitational wave.

2.2 The Michelson–Morley Experiment

One way to determine the distance between widely separated objects is to determine the travel time of light from one to the other and back. This is a technique

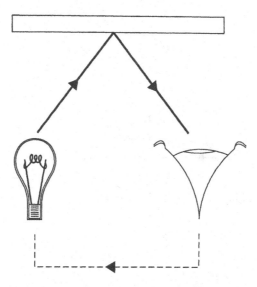

Figure 2.2: A thought-experiment version of a light clock. One "tick" of the clock consists of a flash of light leaving the lamp, travelling to the mirror, then back to the photodetector, which finally signals the lamp to flash again.

for mapping the structure of space-time that embodies the essence of relativity. Every student of physics is familiar with the explanation of relativistic time dilation based on a clock whose ticks consisted of the return of a light beam reflected from a mirror.[12] (See Figure 2.2.) The earliest actual use of light travel time to experimentally determine the structure of space-time is the measurement of Michelson and Morley.[13] This was the experiment that showed that light travelled at a constant speed in all inertial frames of reference. That is not how they described it, of course, since their work predated the theory of relativity. They would have described their experiment as an attempt to see if they could detect an apparent shift in the speed of light due to the Earth's motion through the *ether*.

The experiment was carried out with a device that has come to be called a *Michelson interferometer*. (See Figure 2.3.) In its simplest form, it consists of a source of light, a partially reflecting mirror (the *beam splitter*), and two mirrors located some distance away from the beam splitter in two orthogonal directions. Light transmitted through the beam splitter from the lamp travels toward one mirror, while reflected light travels to the other. The light beams are reflected from each of these mirrors back toward the beam splitter, where they are superposed. The measurement consists of inspecting the light emerging from the *output port* of the beam splitter.

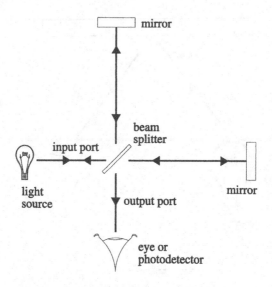

Figure 2.3:   A schematic diagram of the simplest form of Michelson interferometer.

Let's consider some of the features of the Michelson interferometer as a measuring device. The function of this instrument is to compare the amount of time the light beams in the two arms take to complete their paths. Interferometry is well suited for this task. Recall that the light in the two arms comes from a common source. This means that the light enters each arm from the beam splitter in a well defined phase relationship with the light in the other arm. It is simplest to fix attention on one particular phase surface or *wave front* — this wave front enters each arm at the same instant, then propagates toward the far mirrors.

Assume, for the moment, that the two arms have precisely the same round trip length $2L$. If the ether theory were correct, then each beam travels through the ether at the speed of light, c. Since the laboratory is fixed on a moving, spinning Earth, then in most cases light would travel at different speeds in the two arms, as measured with respect to apparatus fixed in the moving laboratory. (It seemed a good working hypothesis in 1887 to assume that the Sun, as one of the "fixed stars", would be at rest with respect to the ether.) The yet-to-be-invented relativity theory would have said instead that light travels at c in all directions as seen in the lab frame, no matter what that frame is. According to the ether theory, when the two wave fronts return to the beam splitter from their travels, they would have taken different amounts of time, and thus will have arrived slightly "out of sync". Or, returning to a picture of a continuous light wave, the two waves returning to the beam splitter will have suffered a *phase shift* with respect to each other. On the other hand, relativity says that if the two arms of the interferometer have precisely the same length, then the two beams will return precisely *in phase*.

The phenomenon of interference between the two returning beams enables one to make a surprisingly precise measurement of the relative phase, or difference in travel times, between the two beams. Let's consider how the state of the light at the output of the interferometer depends on the light travel time of the two beams. We'll follow here the analysis presented in Hermann Haus' *Waves and Fields in Optoelectronics*.[14]

Consider the schematic of a Michelson interferometer shown in Figure 2.3. Call the electric field of the input light $E_0 e^{i(2\pi ft - kx)}$. We can represent the effect of the 50–50 beam splitter by an *amplitude reflection coefficient* $r = 1/\sqrt{2}$, and an *amplitude transmission coefficient* of $t = i/\sqrt{2}$. That is, the light that continues along the $\hat{x}$ axis after passing through the beam splitter has an electric field of $i(E_0/\sqrt{2})e^{i(2\pi ft - k_x x)}$, while the light reflected into the $\hat{y}$ arm has amplitude $(E_0/\sqrt{2})e^{i(2\pi ft - k_y y)}$. We have written different wave numbers k_x and k_y to allow for the putative different speeds of light in the two arms, as proposed by the ether theory.

Reflection at the far mirrors multiplies each wave by -1. When the light in each arm returns to the beam splitter, it again is partly transmitted, partly reflected, with the same coefficients as at the first encounter. The light exiting the interferometer through the output port has the field

$$
\begin{aligned}
E_{out} &= (i/2)E_0 e^{i(2\pi ft - 2k_x L_x)} + (i/2)E_0 e^{i(2\pi ft - 2k_y L_y)} \\
&= (ie^{i2\pi ft}/2)E_0(e^{-i2k_x L_x} + e^{-i2k_y L_y}) \\
&= ie^{i(2\pi ft - k_x L_x - k_y L_y)}E_0 \cos(k_x L_x - k_y L_y).
\end{aligned} \tag{2.10}
$$

Light will also exit through the input port of the beam splitter, as if reflected back toward the lamp, with amplitude

$$
E_{refl} = -ie^{i(2\pi ft - k_x L_x - k_y L_y)}E_0 \sin(k_x L_x - k_y L_y). \tag{2.11}
$$

We can see that the amplitude of the light leaving the interferometer depends on the difference in the phase accumulated by the light travelling in the two arms, $k_x L_x - k_y L_y$. When that difference is zero, E_{refl} vanishes, and light of amplitude E_0 exits from the output port. If the difference were $\pi/2$, all of the light would instead be reflected from the input port, with no light at all leaving through the output port. Since the power in a beam of light is proportional to E^2, we see that

$$
\begin{aligned}
P_{out} &= P_{in}\cos^2(k_x L_x - k_y L_y) \\
&= (P_{in}/2)(1 + \cos 2(k_x L_x - k_y L_y)).
\end{aligned} \tag{2.12}
$$

In other words, a Michelson interferometer acts as a transducer from travel time difference to output optical power. By measuring the brightness of light at the output port, we can learn about the difference in travel time of the two beams, up to an integer number of optical periods $\tau_\lambda \equiv \lambda/c$. We can quickly verify that energy is conserved, by noting that the sum of the power exiting through the output port and the input port is

$$P_{out} + P_{refl} = P_{in}(\cos^2(k_x L_x - k_y L_y) + \sin^2(k_x L_x - k_y L_y))$$
$$= P_{in}. \tag{2.13}$$

We've taken a bit of historical liberty with this last part of the description of the Michelson interferometer. Today, we might set up the Michelson interferometer to operate in just the way described above, that is, as a transducer from travel time difference to brightness. But Michelson and Morley lacked a precise way to measure brightness. So they used a slight variation of the scheme. To see how it worked, recall that the light travels through the interferometer in beams of finite width. When we calculated the amount of light exiting the interferometer, we implicitly assumed that the two beams were precisely aligned; otherwise, if their phase fronts were not parallel to one another, we would have had to describe how the relative phase varied with position across the beam.

Just that extra degree of freedom was what saved the day for Michelson and Morley. They used their interferometer as we described it, except for the fact that the beams were slightly misaligned. Then, a screen placed at an output port shows, not a spot of uniform brightness, but a set of alternating bright and dark bands, known as *fringes* from their resemblance to the fringed border at the end of a piece of woven cloth. A sketch of how the output port might have looked is shown in Figure 2.4. Bright bands appear where constructive interference occurs, dark bands where the interference is destructive. The position of these fringes on the screen is what indicates the relative phase between the two returning beams. You can think of the alternating bands as an interlaced set of bright and dark output ports.

Michelson and Morley chose this deliberate misalignment because it was easier for them to see such fringes shift by a small amount from side to side than it was to estimate whether a spot had brightened or dimmed slightly. The situation is the opposite if one were using photoelectric light detectors for *fringe interrogation*. We will discuss modern techniques below. One bit of jargon has survived from Michelson's day to ours — even when we have well-aligned beams in the output port of an interferometer, we refer to the dependence of the uniform brightness superposition of the two beams on arm length as the *fringe pattern*, or "fringe" for short.

Figure 2.4: An impressionistic rendering of the appearance of interference fringes at the output of a classic Michelson interferometer. Here, the two beams are misaligned as in the Michelson–Morley experiment.

You might wonder how it is possible to set up the interferometer with arms the same length to a precision of a small fraction of a wavelength of light. The answer is that it is hard, although possible, but also not necessary. The interferometer responds in nearly the same way if one arm is shorter than the other by a small integer number of wavelengths. The limit on the tolerable arm length difference is set by the fact that no source of light emits waves of precisely steady wavelength. The arm lengths must match substantially better than the "coherence length" of the light source. Michelson and Morley explain in their paper how they set the arm lengths to be sufficiently close to one another.

But if the prediction of the ether theory is a particular phase difference between the light in the two arms, didn't Michelson and Morley need to start with the interferometer arms set at precisely the same length in order to make a well-defined measurement? No. Regardless of the exact value of the length L of the arms, the ether theory predicts that for an interferometer moving through the ether at speed $v = \beta_\oplus c$ the travel-time difference should be modulated by the amount $2L\beta_\oplus^2/c$ as the interferometer is rotated by 90°. So the amplitude of the modulation is insensitive to small departures of the arm lengths from equality. That is why Michelson and Morley invested much effort in a way to rotate their interferometer gently and smoothly. Looking for a shift in the fringe positions is a natural and well-defined measurement, and does not depend crucially on the exact initial set-up of the interferometer. This is a particular example of a rule of great generality in experimental physics: It is easier to measure an effect you can modulate (or "chop") than to measure a constant effect. We will discuss the advantages of such "relative" experiments over "absolute" ones in greater detail in Chapter 10.

The sensitivity Michelson and Morley were able to achieve is rather impressive. Previously, Michelson had used an interferometer in which light made only a single round trip across the apparatus, but good results were only obtained after additional mirrors were added to lengthen the light path in each arm to total optical length $L_{opt} = 22$ meters. With that value for L_{opt}, the predicted change in travel time difference as the apparatus was rotated through 90° is about 7×10^{-16} sec. Michelson and Morley were able to read the shift in fringe positions to better than $\lambda/20$, equivalent to a time difference of 8×10^{-17} sec.

One of the greatest impediments to the success of this experiment was the sensitivity of the apparatus to externally driven vibrations. This was true despite the fact that the mirrors were securely attached to a massive stone slab floating in a pool of mercury. The best results could only be obtained by working in the wee hours of the morning, when the noise of human activity is at a minimum. The switch to multi-pass arms of greater total length was of no help in making the ether's signal stand up above vibration noise, because multiple reflections increase this kind of noise as much as they do the signal.

2.3   A Schematic Detector of Gravitational Waves

It is indicative of the unity of physics that we find ourselves again turning a century later to the same sort of apparatus, in order to answer another important question about the nature of space-time. Inevitably, many of the same experimental physics issues will arise again. The Michelson–Morley experiment teaches many lessons we need to learn well, since in order to detect gravitational waves we'll have to achieve sensitivity greater by many orders of magnitude.

To detect a gravitational wave, we want to measure the distance between test masses separated by a very large distance. A rigid ruler isn't very practical for measuring the distance, if for no other reason than that a long ruler will not be well approximated as rigid, but will have longitudinal vibration modes of rather low frequency. But we can determine the distance by measuring the round trip travel time of light beams sent over large distances. So it is natural to consider sensing the effect of a gravitational wave on a set of test masses by installing a Michelson inteferometer to monitor their relative separations.

In this section, we'll discuss a schematic experiment to demonstrate how one could detect gravitational waves. Our *gedanken* version of an inteferometric gravitational wave detector looks very much like the Michelson interferometer of 1887. There is one key difference — the mirrors are not connected to a single rigid structure. Instead, each mirror rests on a freely-falling mass, so that it responds in a simple way to gravitational effects.

Since this is a thought experiment, let's consider the simplest possible version: a "50/50" beam splitter sits on one free mass at the vertex of a symmetric "L", while at each end of the L sits another free mass carrying a flat mirror. This works quite a bit like Michelson's original interferometer.

Before we proceed, it is worthwhile to have a brief discussion of the history of this idea. Interferometric detection of gravitational waves is an idea of such beauty that it should not be surprising that it occurred independently to a number of people at roughly the same time. It is also an experiment whose obvious difficulty makes it plausible that initial suggestions were not immediately followed up. An early paper by Gertsenshtein and Pustovoit lays out the key idea, and gives a well-founded, if conservative, estimate of the possible sensitivity.[15] Unpublished work of Joseph Weber and his student Robert Forward in the 1960's formed the groundwork for the construction by Forward and his colleagues of the first operating gravitational wave interferometer.[16] The lineage of the interferometer community active today can be traced to independent work in the early 1970's by Rainer Weiss, who is also credited by Forward with crucial advice on the latter's work. Inspired by Weber's claimed detections of gravitational waves, Weiss conceived of the large interferometers now starting construction, and performed a detailed analysis of their possible performance.[17]

To see how such a gravitational wave detector might work, recall that the Special Theory of Relativity taught us that light's behavior is especially simple — it travels at a constant speed, c, in any inertial frame of reference. The mathematical expression of this idea is that a ray of light connects sets of points separated by an interval of zero, or

$$ds^2 = 0. \tag{2.14}$$

This is the key bit of physics that will enable us to see how a Michelson interferometer can be used to detect a gravitational wave.

We simplify the mathematics, without loss of generality, by imagining that we've laid out our free-mass Michelson interferometer with its arms aligned along the $\hat{x}$ and $\hat{y}$ axes, with the beam-splitter at the origin. Then, since the paths taken by light in each arm will have only either dx or dy non-zero, we need only consider the 11 and 22 components of the metric (along with the trivial 00 component).

First, consider light in the arm along the $\hat{x}$-axis. The interval between two neighboring space-time events linked by the light beam is given by

$$ds^2 = 0 = g_{\mu\nu}dx^\mu dx^\nu$$
$$= (\eta_{\mu\nu} + h_{\mu\nu})dx^\mu dx^\nu$$
$$= -c^2 dt^2 + (1 + h_{11}(2\pi f t - \mathbf{k} \cdot \mathbf{x}))dx^2. \qquad (2.15)$$

This says that the effect of the gravitational wave is to modulate the distance between two neighboring points of fixed coordinate separation dx (as marked, in this gauge, by freely-falling test particles) by a fractional amount h_{11}.

We can evaluate the light travel time from the beam splitter to the end of the $\hat{x}$ arm by integrating the square root of Eq. 2.15

$$\int_0^{\tau_{out}} dt = \frac{1}{c} \int_0^L \sqrt{1 + h_{11}} dx \approx \frac{1}{c} \int_0^L \left(1 + \frac{1}{2}h_{11}(2\pi f t - \mathbf{k} \cdot \mathbf{x})\right) dx,$$

$$(2.16)$$

where, because we will only encounter situations in which $h \ll 1$, we've used the binomial expansion of the square root, and dropped the utterly negligible terms with more than one power of h. We can write a similar equation for the return trip

$$\int_{\tau_{out}}^{\tau_{rt}} dt = -\frac{1}{c} \int_L^0 \left(1 + \frac{1}{2}h_{11}(2\pi f t - \mathbf{k} \cdot \mathbf{x})\right) dx. \qquad (2.17)$$

The total round trip time is thus

$$\tau_{rt} = \frac{2L}{c} + \frac{1}{2c} \int_0^L h_{11}(2\pi f t - \mathbf{k} \cdot \mathbf{x})dx - \frac{1}{2c} \int_L^0 h_{11}(2\pi f t - \mathbf{k} \cdot \mathbf{x})dx.$$

$$(2.18)$$

The integrals are to be evaluated by expressing the arguments as a function just of the position of a particular wavefront (the one that left the beam-splitter at $t = 0$) as it propagates through the apparatus. That is, we should make the substitution $t = x/c$ for the outbound leg, and $t = (2L - x)/c$ for the return leg. Corrections to these relations due to the effect of the gravitational wave itself are negligible.

A similar expression can be written for the light that travels through the $\hat{y}$ arm. The only differences are that it will depend on h_{22} instead of h_{11} and will involve a different substitution for t. The interferometer output will indicate the relative phase shift between the two beams that interfere upon returning to the beam-splitter. That phase shift is simply $2\pi f$ times the difference in travel-time perturbations in the two arms. We could write a general expression for the time travel difference as a function of the gravitational wave's arrival direction, polarization, and frequency, but it is probably more instructive to consider a few special cases. The simplest is for a gravitational wave propagating along the $\hat{z}$ axis. Consider a sinusoidal wave in the + polarization with frequency f_{gw} and amplitude $h_{11} = -h_{22} = h$. If $2\pi f_{gw}\tau_{rt} \ll 1$, then we can treat the metric perturbation as approximately constant during the time any given wavefront is present in the apparatus. There will be equal and opposite perturbations to the light travel time in the two arms. The total travel time difference will therefore be

$$\Delta\tau(t) = h(t)\frac{2L}{c} = h(t)\tau_{rt0}, \tag{2.19}$$

where we have defined $\tau_{rt0} \equiv 2L/c$. We can express this as a phase shift by comparing the travel time difference to the (reduced) period of oscillation of the light, or

$$\Delta\phi(t) = h(t)\tau_{rt0}\frac{2\pi c}{\lambda}. \tag{2.20}$$

Another way to say this is that the phase shift between the light that traveled in the two arms is equal to a fraction h of the total phase a light beam accumulates as it traverses the apparatus. This immediately says that the longer the optical path in the apparatus, the larger will be the phase shift due to the gravitational wave. This is the same scaling rule as in the Michelson-Morley experiment.

But this scaling law won't hold for arbitrarily long arms. Consider the case where the optical path is so long that $2\pi f_{gw}\tau_{rt} \ll 1$ is no longer valid. In contrast to the Michelson-Morley experiment, there is a maximum useful τ_{rt}, as can be seen by considering the case when $f_{gw}\tau_{rt} = 1$. Then the light spends exactly one gravitational wave period in the apparatus; for every part of its path for which the light "sees" a positive value of h, there is an equal part for which it sees an equal but opposite value of h. Thus, in that case there would be no net modulation of the total travel time and thus zero measurable output at the beam-splitter.

To find the response of an interferometer for arbitrary values of τ_{rt}, we should replace our assumption that the gravitational wave's value of h is approximately constant during τ_{rt} by writing $h(t) = h\exp(i2\pi f_{gw}t)$. Then the integral 2.16 that starts at $t = 0$ will be equal to

$$\int_0^{\tau_{out}} dt = \frac{L}{c} + \frac{h}{4\pi i f_{gw}}[e^{i2\pi f_{gw}L/c} - 1], \tag{2.21}$$

while Eq. 2.17 will yield

$$\int_{\tau_{out}}^{\tau_{rt}} dt = \frac{L}{c} + \frac{h}{4\pi i f_{gw}}e^{i2\pi f_{gw}2L/c}[1 - e^{-i2\pi f_{gw}L/c}]. \tag{2.22}$$

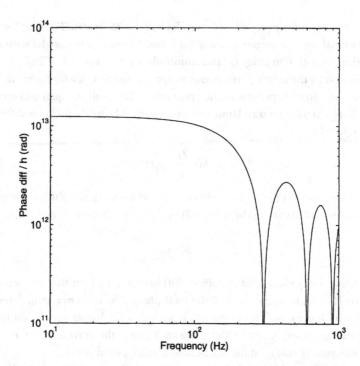

Figure 2.5: The transfer function of a simple Michelson interferometer. In the example, the mirrors that define the ends of the arms are at the rather large distance of 500 km from the beam splitter. The light has a wavelength of 0.5 microns.

As before, we need to also evaluate the effect on light travelling in the $\hat{y}$ arm, and form the difference. We find $\Delta\tau = h\tau_{rt0}\exp(i\pi f_{gw}\tau_{rt0})\,\text{sinc}\,(\pi f_{gw}\tau_{rt0})$, or

$$\Delta\phi(t) = h(t)\tau_{rt0}\frac{2\pi c}{\lambda}\text{sinc}(f_{gw}\tau_{rt0})e^{i\pi f_{gw}\tau_{rt0}}. \qquad (2.23)$$

Here, we make use of the function sinc $x \equiv (1/\pi x)\sin \pi x$. (Note that many authors use a slightly different definition, sinc $x \equiv (1/x)\sin x$.) A graph of the magnitude of $\Delta\phi$ as a function of f_{gw} is shown in Figure 2.5.

A straightforward manipulation of Euler angles gives the angular dependence of the sensitivity of the interferometer, in the low frequency limit. The idea is to determine the elements h_{11} and h_{22} in the $\hat{x} - \hat{y}$ plane, given a description of the wave in a coordinate system whose $\hat{z}'$ axis is aligned with the propagation vector of the wave. One finds (see for example Schutz and Tinto,[18] who followed Forward[16])

$$h_{11}(t) = h(t)[\cos 2\Phi(\cos^2\Psi - \sin^2\Psi\cos^2\Theta) - \sin 2\Phi\sin 2\Psi\cos\Theta] \qquad (2.24)$$

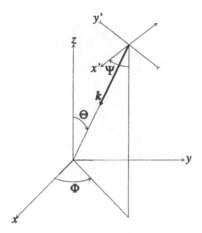

Figure 2.6: Coordinates used to describe the antenna pattern of a gravitational wave detector.

and

$$h_{22}(t) = h(t)[\cos 2\Phi(\sin^2 \Psi - \cos^2 \Psi \cos^2 \Theta) + \sin 2\Phi \sin 2\Psi \cos \Theta]. \quad (2.25)$$

(The angles are defined in Figure 2.6.) One now carries through the integrals along the $\hat{x}$ and $\hat{y}$ arms as we did above. But now, instead of $h_{11}/h_{22} = -1$, the two components may have an arbitrary ratio. We can skip ahead to the answer by recognizing that the whole previous calculation goes through as before, so long as we substitute for $h(t)$

$$h \rightarrow \frac{1}{2}[h_{11}(\Theta, \Phi, \Psi) - h_{22}(\Theta, \Phi, \Psi)], \quad (2.26)$$

the "average" appropriate to our geometry. Then we find

$$\Delta\phi(t) = h(t)\tau_{rt0}\frac{2\pi c}{\lambda} \left(\frac{1}{2}(1 + \cos^2 \Theta) \cos 2\Phi \cos 2\Psi - \cos \Theta \sin 2\Phi \sin 2\Psi \right). \quad (2.27)$$

This expression reveals a response as a function of angle, or *beam pattern*, that is remarkably non-directional. (See Figure 2.7.) Waves propagating along the $\hat{x}$ and $\hat{y}$ axes (with the proper polarization) show response down only by a factor of 2 from the maximum $\hat{z}$ axis response. The only true nulls are along the bisector between the $\hat{x}$ and $\hat{y}$ axes; in these directions the interferometer arm length changes are always equal in the two arms, giving no response. There are also nulls, in the limit of low

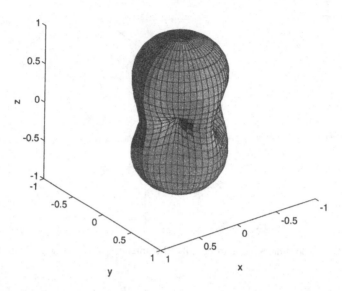

Figure 2.7: The sensitivity, as a function of direction, of an interferometric gravitational wave detector to unpolarized gravitational waves. The interferometer arms are oriented along the x and y axes.

frequency, in the orthogonal directions in the $\hat{x} - \hat{y}$ plane.[19] The rms average of this sensitivity pattern over all angles is $1/\sqrt{5} \approx 0.45$.

The broad angular response of a gravitational wave inteferometer is both a blessing and a curse for gravitational wave astronomy. An interferometer behaves nothing like a telescope that one points at an interesting portion of the sky. It is more like an ear placed on the ground; waves coming from nearly any direction are registered. On the one hand, this is an advantage for surveying the sky, especially if one fears there might be few signals strong enough to be detected. On the other hand, understanding a source of gravitational waves as an astronomical object would get powerful assistance from information acquired with optical telescopes or other receivers of electromagnetic radiation. Identifying the source on the sky will require a rather accurate determination of the position of the source on the sky. That information will have to be extracted from determination of the differences in arrival times of signals at detectors at widely separated locations. We will discuss position determinations in more detail in Chapter 15.

The expression derived above for the beam pattern needs refinement at high frequencies. The reason is that once the light travel time becomes comparable to or larger than the gravitational wave period, the simplifying assumption that $h(t)$ is constant during a wave front's trip is no longer valid. Instead, the total phase shift accumulated by a wave front depends on a detailed consideration of the value of h_{11} or h_{22} as a function of time at the location of the optical wave front as it travels through the interferometer. The method we sketched out above in Eq. 2.18 remains applicable, but the algebra becomes more involved. The results also depend in detail on whether, and if so how, the light makes multiple trips through each arm. We will consider the transfer function of such folded interferometers in Chapter 6.

2.4 Description of Gravitational Waves in Terms of Force

In most respects, gravity in the laboratory can still be considered basically "a push or a pull",[20] even when we take into account General Relativity. Treating gravitational waves as a quasi-Newtonian phenomenon gives them a comfortable familiarity. It is also by far the simplest picture in which to include other "true" forces, such as the electrostatic forces that make a collection of atoms behave like an elastic solid. To take this into account, we need to use a slightly different way of defining coordinates than is given by the transverse traceless gauge. What corresponds most to our usual Newtonian intuition about experiments is the *proper reference frame* description.[9] Instead of coordinates being marked out by freely-falling masses, in this description of events we mark them by perfectly rigid rods arranged in an orthogonal framework.

Imagine that we have placed two freely-falling test particles of mass m a small distance apart along the $\hat{x}$ axis, at $x = \pm L$. If a gravitational wave is incident along the $\hat{z}$-axis in the $+$-polarization, how does the system respond? We saw above that if we used the travel time of a light beam to measure how the distance changed, then the distance change is equal to the distance times $h_{11}/2$.

Measuring distances with rigid rulers ought to give us the same result. But if we define coordinates as points that are fixed along a rigid ruler, then we can only have the separation between masses changing if their coordinates change, as seen in such a frame. Thus, in such a description of physics, we need to describe a gravitational wave as something that can accelerate a freely-falling test mass. In other words, we need to describe a gravitational wave in terms of a force. It is not hard to show that we would get the same answers as before if we took as the *tidal force* due to a gravitational wave

$$F_{gw} = m\ddot{x} = \frac{1}{2}mL\frac{\partial^2 h_{11}}{\partial t^2}. \tag{2.28}$$

This is a perfectly acceptable description as long as separation of the test masses is small compared to the wavelength of the gravitational wave. (Recall that if we

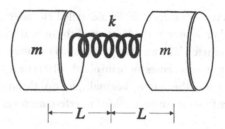

Figure 2.8: A thought-experiment version of a resonant mass detector of gravitational waves, or "bar".

integrate the effect on an optical wavefront through a longer distance, the gravitational wave will have time to change sign, and reduce the net observable effect.) This is the only straightforward way to include other forces in the problem.

As an example, a resonant mass detector can be modelled as a pair of masses connected by a spring. (See Figure 2.8.) We would analyze such a detector's interaction with a gravitational wave as we did above, with the single addition of the inclusion of a term for the spring force in the equation of motion. There are various ways to detect the relative motion of the ends of the bar. Weber's original scheme was to equip the middle of the bar (the "spring") with piezoelectric strain gauges.[3] Many of Weber's successors have instead placed an inertial sensor (i.e., an accelerometer) on one end of the bar. We'll discuss resonant-mass detectors in more detail in Chapter 13.

It makes a difference to the words we use to describe gravitational effects whether we choose to use this proper reference frame picture, or whether instead we use the TT gauge and define points in our coordinate system by the positions of freely-falling test masses. In the proper reference frame, gravitation appears as a force that accelerates "free" test masses with respect to rigid rods. In the TT gauge, by contrast, there is no change in the coordinates of a freely-falling mass. The effect of gravitation shows up instead as a change in the metric describing the space between the non-moving test masses. But in spite of the differences in language, the two pictures give completely consistent descriptions of the results of experiments, as long as we are in the short distance regime where both are equally applicable. Whenever we need to consider systems comparable to or longer than a gravitational wavelength, then the TT gauge picture is the only sensible one to use.

Sources of Gravitational Waves 3

We start this chapter with a discussion of the physics involved in the generation of gravitational waves. The remainder of the chapter is devoted to a review of the possible astrophysical sources of gravitational radiation, and of the characteristics of the signals they would produce.

3.1 Physics of Gravitational Wave Generation

The existence of gravitational waves would be of only formal interest if there were no way to generate them. There are, of course, mechanisms for doing just that. The physical intuition that we physicists have learned from the study of electromagnetic waves is applicable, at least by analogy, to the gravitational case. Just as in electro dynamics, the emission of gravitational waves can be expressed exactly in terms of a retarded potential. (See, for this and many other results in this section, the treatment in any relativity text, for example Misner, Thorne, and Wheeler, Chapter 36.[11]) But again, just as in electrodynamics, in most practical problems it is much simpler to work with the approximation known as the multipole expansion, good in the limit that the size of the source is much smaller than the wavelength, $r_{source}/\lambda \ll 1$.

Recall that in the typical electromagnetic radiation problem, the dominant contribution to the radiated field comes from the time variation of the electric dipole moment, as expressed by

$$\mathbf{E} = \frac{1}{Rc^2}(\ddot{\mathbf{d}} \times \mathbf{n}) \times \mathbf{n}. \tag{3.1}$$

In this equation, R is the distance from the source to the observation point, $\mathbf{n}$ is the unit vector pointing from the source to the observer, and $\mathbf{d}$ is the electric dipole moment, defined as

$$\mathbf{d} \equiv \int dV \, \rho_q(\mathbf{r})\mathbf{r}. \tag{3.2}$$

Here, ρ_q is the charge density and the integral runs over the volume of the source. Fields radiated by the next highest moments, the magnetic dipole and the electric quadrupole, are weaker by an additional power of r_{source}/λ.

Why is there no electromagnetic monopole radiation? For such a term, we would require a time variation in the "monopole moment", that is to say in the total electric charge of the source. But since no change in the total charge of an isolated system is allowed, monopole radiation is forbidden. The most that can happen is that charges can move from one part of the isolated source to another; these motions give rise to the dipole and higher moments in the expansion.

The great similarity between Newton's Law of Gravitation and Coulomb's Law gives reason to hope that deep analogies will remain in their relativistic generalizations, general relativity and the Maxwell theory. Of course, there are well-known crucial differences as well, even at the pre-relativistic level. Among the most important is the fact that there are two possible signs to the electric charge, while mass, the "charge" of gravity, only appears with one sign. Furthermore, the gravitational charge is, by the Principle of Equivalence, also the measure of the inertia of a body. These features bring additional conservation laws into play in the radiation process.

Conservation of energy plays the same role for gravitational radiation as does charge conservation in the electromagnetic case. So there is no monopole term for gravitational radiation either. What about dipole moments? One can define a gravitational dipole moment as

$$\mathbf{d}_g \equiv \int dV \, \rho(\mathbf{r})\mathbf{r}, \tag{3.3}$$

with $\rho(\mathbf{r})$ the mass density, and $\mathbf{r}$ measured with respect to any origin of our choosing. But the law of conservation of momentum requires that $\dot{\mathbf{d}}_g$ remain a constant for any isolated system. Without a possibility of higher time derivatives, there can be no radiation associated with this moment of the mass distribution.

We could also define a gravitational analog of the magnetic dipole moment

$$\mu_g \equiv \int dV \, \rho(\mathbf{r})\mathbf{r} \times \vec{v}(\mathbf{r}). \tag{3.4}$$

But this must remain constant, because of the law of conservation of angular momentum. So at this level as well, no gravitational waves are generated.

One might almost begin to despair that there is a conspiracy in nature to forbid gravitational radiation altogether. But in fact we have run out of relevant conservation laws. Higher moments of mass distributions may vary, and will in fact generate gravitational waves.

In the typical case, where motions within the source are slow compared with the speed of light, it is the time variation of the gravitational quadrupole moment that contributes most strongly. We can define the *reduced quadrupole moment* as

$$I_{\mu\nu} \equiv \int dV \left(x_\mu x_\nu - \frac{1}{3}\delta_{\mu\nu} r^2 \right) \rho(\mathbf{r}). \qquad (3.5)$$

(It is called "reduced" because this definition is smaller by constant factors than some other popular definitions of quadrupole moments.) Such a calculation is easiest to picture when gravity within the source is weak, so that we can treat the motions as occurring within flat space-time. But the moment remains well-defined even in the strong gravity case — it could always be measured, in principle, by a (not too) distant observer, as the coefficient of the quadrupole term in the quasi-Newtonian expansion of the gravitational potential.

We can now write the gravitational analog of Eq. 3.1, for the strongest allowed component of gravitational radiation. It is

$$h_{\mu\nu} = \frac{2G}{Rc^4} \ddot{I}_{\mu\nu}, \qquad (3.6)$$

where the right hand side is to be evaluated at the retarded time $t - R/c$.

It is a useful exercise to actually work out the moments in a particular case. We examine the waves emitted by a pair of equal point masses moving in a circular orbit about their common center of mass. (See Figure 3.1.) This example is a simple model of a binary star system, one of the most common astronomical sources of gravitational waves.

Assume each object has a mass of M, and that the total distance separating them is $2r_0$. The assumption of circular motion implies a constant orbital frequency, called f_{orb}. Call the direction of the normal to the orbit plane the $\hat{z}$ direction. Then,

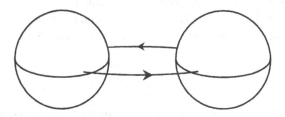

Figure 3.1: A schematic representation of a binary star system. The two stars orbit about the pair's center of mass.

it follows directly from Eq. 3.5 that the components of the quadrupole moment are

$$I_{xx} = 2Mr_0^2 \left(\cos^2 2\pi f_{orb} t - \frac{1}{3} \right), \tag{3.7}$$

$$I_{yy} = 2Mr_0^2 \left(\sin^2 2\pi f_{orb} t - \frac{1}{3} \right), \tag{3.8}$$

and

$$I_{xy} = I_{yx} = 2Mr_0^2 \cos 2\pi f_{orb} t \sin 2\pi f_{orb} t. \tag{3.9}$$

The components involving z are uninteresting: $I_{zz} = -\frac{1}{3}Mr_0^2$, a constant, while crossterms of z with x and y vanish.

It is straightforward to calculate the second time derivatives, $\ddot{I}_{\mu\nu}$. If, for example, we are interested in the gravitational wave amplitude at a point along the z axis at a distance R, we can simply plug in to Eq. 3.6 above to find

$$h_{xx} = -h_{yy} = \frac{32\pi^2 G}{Rc^4} Mr_0^2 f_{orb}^2 \cos 2(2\pi f_{orb})t, \tag{3.10}$$

and

$$h_{xy} = h_{yx} = \frac{-32\pi^2 G}{Rc^4} Mr_0^2 f_{orb}^2 \sin 2(2\pi f_{orb})t. \tag{3.11}$$

Since the amplitude of a gravitational wave is a dimensionless quantity, it would be good to see if we can express it in a form more manifestly dimensionless. A little bit of rearranging of Eq. 3.10 or 3.11 will lead us to a nice insight.[21] We can collect the combination $2GM/c^2$ that appears in the prefactor, recognizing it as the Schwarzschild radius r_S of one of the point masses. By symmetry, we might expect that the amplitude ought to be proportional to the product $r_{S1}r_{S2}$ of the Schwarzschild radii of the two masses. The wave amplitude must always be inversely proportional to the distance R from source to observer. Is there another length by which we can divide $r_{S1}r_{S2}/R$, to make an expression that gives the magnitude of Eq. 3.10? Yes. Newtonian mechanics gives a simple expression for the square of the orbital frequency, $f_{orb}^2 = GM/16\pi^2 r_0^3$. Plugging this in and rearranging, it is easy to see that the amplitude of the gravitational wave from a binary can be expressed as

$$|h| \approx \frac{r_{S1}r_{S2}}{r_0 R}. \tag{3.12}$$

Not only is this expression manifestly dimensionless, but by its beauty it suggests how fundamental is the physical mechanism by which gravitational waves are generated.

Later in this chapter we will examine the likely strength of such waves from actual astronomical systems. But to get a sense of the scale of the problem, we can plug in some representative values for a binary. Consider a binary system of two neutron stars, each of whose masses is roughly the Chandrasekhar mass, or 1.4 times the mass of the Sun. That is, let's take $M = 1.4 M_\odot \approx 3 \times 10^{30}$ kg. The two neutron stars can get close enough to almost touch, at about $r_0 = 20$ km, at which point their orbital frequency is about $f_{orb} \approx 400$ Hz. (Astonishingly, these are actually realistic numbers for some objects!) Assume the binary is located in the nearest cluster of galaxies to us, the Virgo Cluster, at $R \approx 15$ Megaparsec (or Mpc) $\approx 4.5 \times 10^{23}$ m. Then we have

$$h \equiv |h_{\mu\nu}| \approx 1 \times 10^{-21}. \tag{3.13}$$

Note that the gravitational waves are emitted at twice the orbital frequency, a signature of their quadrupolar nature. Note also that measurements made along the $\hat{z}$ axis would observe a circularly polarized gravitational wave: $|h_{xy}| = |h_{xx}|$, and the two polarizations are shifted one quarter cycle in phase with respect to each other. Less special orientations will yield elliptical polarization, reducing to linear polarization for observations made in the $\hat{x} - \hat{y}$ plane.

3.2 In the Footsteps of Heinrich Hertz?

The estimate $h \sim 10^{-21}$ given above is rather astonishingly small. It ought to cause us to think twice whether detection of such small effects will ever be possible. (The rest of this book will argue, no surprise, that it should in fact be quite feasible.) It is also uncomfortable for a physicist to have to rely on naturally occurring astronomical systems to test something as basic as one of the fundamental laws of physics. Of course, gravity itself was discovered by Newton through analysis of astronomical observations.

Ideally, though, one would like to construct a source of gravitational waves entirely under one's control. Then amplitude, frequency, and polarization of the waves could be varied at will, and an absolutely clean test of the predictions of general relativity would be possible. The verification of Maxwell's 1864 prediction of the existence and nature of electromagnetic waves came in just such a series of experiments, carried out by Heinrich Hertz in the late 1880's. A nice discussion of the history of this episode is given in Malcolm Longair's book *Theoretical Concepts in Physics*.[22] An excellent account of the experimental details may be found in the catalog of a museum exhibit of reproductions of Hertz's apparatus.[23]

There was already reasonably good reason to believe in the existence of electromagnetic waves. That the speed of light should match Maxwell's prediction for the speed of electromagnetic waves was remarkable, given that the latter was based on purely electrical measurements, made at zero frequency (or "DC" in experimenters' jargon), of the dielectric constant and magnetic permittivity of free space. Sense could also be made of the wave properties of light, as displayed by interference phenomena, as well as of its polarization. But there was some reason to despair whether direct confirmation would ever be possible. Recognizing the importance of experimental demonstration of the theoretical predictions, the Prussian Academy of Sciences offered in 1879 a prize for experimental proof of the Maxwell theory.

It was not until nine years later, some twenty-four years after Maxwell's first paper on the subject, that such proof was published. (By then, the prize offer had expired!) Hertz faced the daunting challenge of inventing both a suitable controlled source of electromagnetic waves and a detector of sufficient sensitivity. It is easy today to overlook the difficulty this presented, living as we do in a world in which electronic amplification of weak electrical signals is ubiquitous. The technology inside a $10 transistor radio set was utterly beyond Hertz's reach. He was limited to strictly passive devices.

Hertz solved the problem by constructing resonant circuits for the transmitter and receiver, choosing of course to tune them to the same frequency. He used an induction coil and mercury switch to generate a large voltage pulse from the energy stored in a battery. This was applied to a resonant dipole, consisting of a pair of straight wires (inductors) with large metal spheres (capacitors) at their far ends. The near ends of the wires were close together, forming a spark gap to which the input pulse was applied. The pulse excited an electrical oscillation of the tuned circuit, damped rather strongly by the emission of electromagnetic waves.

A second resonant circuit, consisting of a simple loop of wire and a spark gap tuned to the same resonant frequency, was placed some distance away. It was excited into oscillation by the burst of radiation emitted by the transmitter. In what is perhaps the cleverest aspect of the experiment, the resonant excitation was made visible by the creation of sparks in the gap left in the loop of wire. The induced electromotive force in the receiving coil caused a large enough electric field in the gap to ionize the air there. The peak output power of Hertz's transmitter was about 16 kW.

The faint glow of light in the gap, seen in temporal coincidence with the discharge of the transmitter, revealed finally the electromagnetic nature of light. It also represented the world's first microwave communications link, albeit one of small range, minuscule bandwidth, and negligible utility.

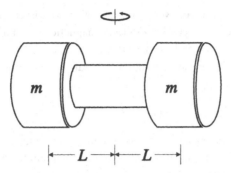

Figure 3.2: A schematic diagram of a rotating dumbbell, a thought-experiment version of a laboratory generator of gravitational waves.

In quick succession, a set of further experiments demonstrated the polarization of the waves, showed that they obeyed the laws of reflection and refraction, could be focused, formed standing waves, etc. This work instantly transformed Maxwell's theory from a rather ghostly and abstract idea to one of great physical immediacy and, eventually, practical application. This story would be worth retelling simply as an example of scientific cleverness and perseverence. But it also represents a rather faithful model for what one would like to accomplish for gravitational waves. That it seems impossible a century after Hertz's work, in spite of the great progress in measurement technology over the intervening years, is a graphic demonstration of the vast gulf in strength that separates the gravitational interaction from its electromagnetic analog.

Let's carry out a sample calculation for a laboratory version of the binary star system that we considered above. (See Figure 3.2.) Assume we could construct a dumbbell consisting of two masses of 1 tonne each, at either end of a rod 2 meters long. Spin this quadrupole about an axis orthogonal to the connecting rod passing through its midpoint, at an angular frequency $f_{rot} = 1$ kHz. Neglecting for simplicity the contribution of the connecting rod, we have a system described by the same equations as the binary star system, Eqs. 3.10 and 3.11. The amplitude of the gravitational waves generated by this device will be

$$h_{lab} = 2.6 \times 10^{-33} \, \text{m} \times \frac{1}{R}. \tag{3.14}$$

Before we rush to plug in a distance R of a few meters, as Hertz was able to do for his experiment, we need to remember that wave phenomena are only distinguishable from near-field effects in the "wave zone", that is at distances from the source comparable to or larger than one wavelength. With $\omega_{rot} = 2\pi \times 1$ kHz, we have $\lambda = 300$ km! The receiver for our Hertzian experiment must be at least that far away from the transmitter. Hertz's electromagnetic experiments involved waves of 6 meters down to 60 cm in length, so the distance across the lab was fine for him.

At a distance of one wavelength, our laboratory generator gives gravitational waves of amplitude

$$h_{lab} = 9 \times 10^{-39}. \tag{3.15}$$

Far from solving the problem of weak astronomical sources, we haven't been able to come within 16 orders of magnitude of matching their amplitudes at the Earth.

Even creating such a strong source as this may not be practicable. Consider the stress in the connecting rod of the dumbbell. It must supply the centripetal force necessary for the masses to move in a circle. If the rod were made of good steel, it would need a cross-sectional area substantially greater than that of a 1 tonne sphere in order not to fail under the stresses in a device with the parameters we have assumed. So we'd have to reduce the rotation frequency to keep the generator from flying apart, with a consequent reduction in the transmitted wave amplitude.

Thus, it appears that it is quite impracticable to construct our own laboratory source of gravitational waves. Gravitational wave emission from the particle beams in high energy accelerators,[24] as well as from nuclear explosions,[25] have also been evaluated, and found wanting. It is for this reason that experimental study of gravitational waves is directed toward waves of astronomical origin. They are simply, and by far, the strongest gravitational waves we are likely to encounter.

3.3   Observation of Gravitational Wave Emission

Astronomy has, in fact, already provided solid evidence for the existence of gravitational waves. This comes from the study of a binary system very much like the example discussed just above, the so-called Binary Pulsar, PSR1913+16. The observations of this system constitute one half of a Hertzian experiment for gravitational waves, as well as an example of the remarkable precision of late 20th century measurement technology.

The pulsar PSR1913+16 was discovered in 1974 by Russell Hulse and Joseph Taylor.[26] These radio astronomers were searching for new examples of pulsars, rotating neutron stars that emit brief bursts of radio waves once or twice per rotation period. Such objects are of great interest in their own right, and we'll discuss below several different circumstances in which they are associated with the emission of strong gravitational waves. For the purpose of the immediate discussion here, though, we need only recognize a few of their properties: they are massive (typically

$M = 1.4M_\odot \approx 3 \times 10^{30}$ kg), nearly point-like (radius of about 10 km), and carry an excellent clock (their rotation rate) that we can read here on Earth by counting the radio pulses that they emit.

Soon after its discovery, PSR1913+16 revealed itself to belong to an especially interesting system. Its pulse rate varied over a dramatic range, with a modulation period of about 7.75 hours. The regularity of this variation was easily explained as being due to the Doppler shift of a steady pulsar in orbit about the center of mass of a binary system. Subsequent observations, pursued by Taylor and his colleagues,[27] have been able to determine nearly all of the relevant properties of this binary system. The other object is also a neutron star, though not visible as a pulsar. The two follow an orbit with an eccentricity of 0.6171, moving at a maximum speed of roughly 400 km/sec.

The determination of these facts, and others about the system, represents a *tour de force* of relativistic physics. If classical physics holds, a binary system with just one visible component, whose motion is observed only via the Doppler effect (a so-called *single-line spectroscopic binary*), cannot have all of its orbital parameters determined unambiguously. In particular, it is typically impossible to disentangle the stellar masses from the inclination angle of the binary orbit.[28] But in no other type of system can the ticks of a clock on one of the stars be timed to a precision of 50 μs over a span of nearly 20 years. This, and the extreme parameters of the system itself, has allowed the full panoply of general relativistic effects to be observed, including orbital precession, gravitational redshift, and radiation time delay (the Shapiro effect). These effects depend on different combination of the orbital parameters, thus allowing unambiguous description of the binary's properties.

It took good fortune to have discovered PSR1913+16 and insight to recognize how interesting it would be as a test bed for general relativity. To make the ultra precise series of observations, continuing from the discovery to the present day, has also required hard work and great ingenuity. A special purpose low-noise wide band receiver is used at the world's largest radio dish, the Arecibo telescope in Puerto Rico. Custom circuitry accurately superposes the signals of various radio frequencies that have become separated by dispersion in the interstellar medium. State-of-the-art atomic clocks provide the most reliable terrestrial timebase. A massive computer program constructs the best fit to the mountains of data in order to expose the various relativistic effects necessary to model the binary system.

For all of the beautiful physics involved in the study of PSR1913+16, indisputably the most exciting thing to come from it is direct observation of an effect that can only be explained if the system is emitting gravitational waves at the rate predicted by general relativity. Since its discovery, the pulsar has been gradually creeping ahead of where it would be in its orbit if it had a constant period. By now, this orbital phase shift amounts to about 10 seconds over 15 years, growing quadratically in time. This can't be due to a mistaken determination of the orbital period, as that error would give a discrepancy that grew linearly in time. The cause must be a loss of orbital energy from the system, causing the neutron stars to fall toward each other, and thus to orbit at a more rapid rate.

Gravitational waves carry energy, just as their electromagnetic counterparts do. A source of gravitational quadrupole radiation has a total *gravitational luminosity* $\mathcal{L}$ given by

$$\mathcal{L} = \frac{G}{5c^5} \langle \dddot{I}_{\mu\nu} \dddot{I}^{\mu\nu} \rangle. \tag{3.16}$$

(The repeated index summation convention applies here, so we square the third time derivatives of all of the components of the quadrupole moment. The angle brackets represent a time average over several periods of the wave.) This energy must come from somewhere. In the case of a binary system, it is drawn from the orbital energy. As the waves carry away energy, the separation of the stars must shrink (recall the Newtonian expression $E_{tot} = -GM_1 M_2 / 2r$), and the orbital motion will consequently speed up.

The Binary Pulsar has a rather eccentric orbit, $e = 0.617$. This means that higher harmonics of the orbital frequency than the second contribute substantially to the luminosity.[29] Nevertheless, the radiation is precisely predicted by the theory of general relativity.[30] When the predicted orbital phase shift is matched with that experimentally determined in PSR1913+16, the values match to about 1%, the experimental accuracy.

It is impossible to stress too much the beauty and importance of this result. Gravitational waves exist, and are emitted at the rate predicted by general relativity. In recognition of the tremendous importance of this discovery, Hulse and Taylor were awarded the Nobel Prize for Physics in 1993.

Does this mean we should declare victory and go home? No. There is still much to be learned about gravitational waves, and much to be learned with them. To complete a basic program of fundamental physical tests of gravitational wave properties, we need to see that they cause the predicted effects on test masses. We'd also like to measure their speed and their polarization.

The present state of gravitational wave physics is in some ways analogous to the case of neutrino physics in the 1940's and 50's. The puzzling continuous spectrum of β-decay energies led Pauli to propose in 1930 that an unseen particle must be carrying away energy and momentum.[31] But in spite of the elegant way in which it resolved a number of problems in physics, it remained for years a ghostly hypothesis.

One contemporary description of the situation had it that "There can be no two opinions about the practical utility of the neutrino hypothesis ... but ... until clear experimental evidence for the existence of the neutrino could be obtained ... the neutrino must remain purely hypothetical."[32] So it was undeniably a great advance when direct observation of the ν itself was achieved in 1959 by Reines and Cowan.[33] It has of course subsequently proven to be a valuable probe of the structure of matter. And, more recently, observations of neutrinos emitted by the Sun and from Supernova 1987A have turned it into a unique tool for astronomy as well.

One of the greatest opportunities provided by the study of gravitational waves is the chance to learn more about the sorts of astronomical objects that emit them. Adding radio, infrared, UV, X-, and γ-radiation to the astronomer's palette has allowed us to paint an ever more vivid and complete picture of the Universe. How much more might we see if we add, not just another color of the same information carrier, but a whole new class of wave? We'll start the consideration of that question in the next part of this chapter, and return to it again in Chapter 15.

3.4 Astronomical Sources of Gravitational Waves

A variety of objects and processes in the Universe may give rise to detectable gravitational waves here on Earth. The list includes mainly, though not exclusively, so-called *compact objects*, specifically neutron stars and black holes. These are among some of the most interesting objects in astronomy, about which much remains to be learned. The "flip side" of this is a fair amount of uncertainty at present about what to expect.

We present here a survey of the conventional wisdom, and some speculation, about what sorts of gravitational waves ought to be arriving at our detectors. We will concentrate on sources that should produce waves in a frequency band accessible to Earth-bound instruments. For now, let us just assume the accessible band to span the range between a few Hz and and a few kHz — the task of understanding and justifying that claim will occupy the middle third of this book.

The chain of reasoning used in making predictions about possible gravitational wave signal strengths takes the following form: The astrophysicist considers a kind of source, and from a study of its dynamics is able to calculate the history of its quadrupole moment. From this follows the strength and temporal or spectral character of the gravitational radiation that it emits. Often it is the case that the difficulty of this task lies less in the subtleties of general relativity than in the complexity of astronomical systems.

In addition, it is necessary to estimate the rate of occurrence of such sources in some fiducial volume, such as a galaxy. For impulsive sources, one calculates a maximum likely signal strength by assuming that the source is located on the surface of a sphere centered on the detector large enough to contain one event per month, or year, or however long an experimenter is able to wait. For periodic sources, one assumes the source is approximately $n^{-1/3}$ away from the antenna, where n is the number density of sources. Of course, for known individual objects one uses the true distance away. The duration and frequency content determine to some extent the experimenter's design of his apparatus and data analysis strategy. The expected signal strength, when compared with instrument noise levels, indicates whether it is likely that a successful detection can be made.

3.4.1 *Neutron star binaries*

Binary stars are among the commonest astronomical systems. Calculation of their gravitational radiation is quite simple, as we saw above. Every known binary involving luminous stars, though, has such a long period that the gravitational wave signal is far outside the passband of any terrestrial antenna. One of the shortest periods known is that of WZ Sge, 81 minutes. Sources with such long periods could only be observed by an antenna in space, free of the seismic disturbances of the terrestrial environment. (See Chapter 9 for a comparison of terrestrial and space-borne detectors.) It is unfortunate that these sources will remain beyond our reach for some time, since by their very simplicity they allow the cleanest possible test of the general relativistic prediction of the strength and polarization of gravitational radiation.

The system we discussed in the previous section, PSR1913+16, also suffers from the same defect. The gravitational waves it emits now are at $f \approx (4 \text{ hr})^{-1}$ and a few low order harmonics. But as its orbit continues to decay, the Binary Pulsar will gradually transform itself into a source of high frequency gravitational waves. For this reason, it serves as a prototype of one of the most important sources of gravitational waves for terrestrial detectors. Over an interval of about 10^8 years, PSR1913+16's orbit will become circular, and the two neutron stars will gradually approach each other as they orbit at an ever-increasing frequency, until they are within a few stellar radii of each other. In a final blaze of glory, the pair will emit a strong burst of gravitational radiation as the two stars spiral toward mutual assured destruction. In the last moments of such a system, its parameters will in fact be a close match to the numbers we used in Eq. 3.13 above.

Exact calculation of the waveform in the final death plunge of a neutron star binary is something of a challenge, since it involves relativistic effects in a strong way, as well as knowledge of the equation of state of the neutron stars. But up to those final milliseconds, the history of $\ddot{I}$, and thus of h, can be calculated with reasonable accuracy using the weak field approximations that we made above.

The resulting *chirp* waveform from a pair of neutron stars, quasi-sinusoidal but sweeping up in frequency and in amplitude, has been studied by several authors. It is possible to write down simple expressions for the history of the signal frequency $f(t)$ and amplitude $h(t)$ as long as the weak field approximations hold.[34] The wave parameters are

$$f(t) \approx 2.1 \text{Hz} \times \left(\frac{M_1 + M_2}{M_1^3 M_2^3} \right)^{1/8} \left(\frac{1 \text{ day}}{\tau} \right)^{3/8}, \qquad (3.17)$$

and

$$h(t) \approx 6.6 \times 10^{-24} \left(\frac{15 \text{ Mpc}}{R} \right) \left(\frac{M_1^3 M_2^3}{M_1 + M_2} \right)^{1/4}$$

$$\times \left(\frac{1 \text{ day}}{\tau} \right)^{1/4} (1 + 6 \cos^2 \theta + \cos^4 \theta)^{1/2}, \qquad (3.18)$$

where M_1 and M_2 are the masses of the two neutron stars in units of the solar mass, and τ is the time until collision. The angle θ is the inclination of the orbit of the system to the line of sight, and R is its distance from us. Before collision or disruption, the system is likely to reach frequencies of almost 1 kHz, and amplitudes near 10^{-21} for a system in the Virgo Cluster of galaxies, about 15 Mpc distant. This corresponds to several percent of the rest mass radiated away, a rather high efficiency.

The question of the rate of such decays was originally addressed by Clark, van den Heuvel, and Sutantyo.[35] More reliable estimates were made recently.[36] According to the best present conservative estimates, we must see to distance of 200 Mpc in order to expect to see 3 events per year. At such a distance, the maximum amplitude is just below $h \approx 10^{-22}$.

3.4.2 *Supernovae*

Supernovae are among the most dramatic events seen in the sky. In the past, the appearance of such a bright "new star" was considered an ill omen. The great 16th century astronomer Tycho was inspired to devote his life to the study of the heavens when as a young man he observed a supernova, surely a good omen for us. In the intervening years, astronomers observed many supernovae in external galaxies. The phenomenon of the sudden drastic brightening of a star and its subsequent gradual decay was observed in two basic patterns, called Type I and Type II. (Astronomers have a way with words.) The kind called Type I is thought to involve the explosive detonation of a white dwarf star, without substantial emission of gravitational waves.

But the supernovae of Type II may emit strong gravitational waves. A good review of our pre-1987 knowledge of supernovae was given by Trimble.[37]

Since the time of Tycho, it took until early 1987 for another supernova nearby enough to observe with the naked eye to appear. Supernova 1987A, as it was called, was caught early in the act of brightening, and became the best studied supernova of all time. The significance of 1987A was that it allowed a detailed check of the predicted mechanism of the Type II supernova event. The key features of the model were shown to be correct.[38]

The mechanism for gravitational radiation from a supernova may be summarized as follows: A massive star produces a core of roughly 1.4 $M_\odot$ which has burned to iron. Such a core, when seen without the surrounding layers, would be called a *white dwarf* star. Above this critical *Chandrasekhar mass*, the electron degeneracy pressure which has supported the core can no longer resist the weight of the star. The matter in the core suddenly is converted into neutrons. Without sufficient pressure to support its weight, the stellar core begins to collapse. This collapse is reversed in a sudden *bounce* once the core has reached nuclear densities, around 3×10^{14} gm/cm^3. If the collapse is not spherically symmetric, then the bounce is responsible for a large value of $\dddot{I}$ and hence a burst of gravitational radiation. Depending on the degree of damping the first bounce may be followed by several others, accompanied by further strong emission of gravitational waves.

Meanwhile, the energy that is released in the collapse is deposited in the outer layers of the star. This energy comes mainly in the form of the neutrinos created from the conversion of the core to neutrons. This results in the energetic explosion of those outer layers, including the brilliant emission of light that defines the process. Interestingly, the physics of this explosion has proven to be challenging to understand in detail.[39] But when all of the smoke has cleared, a neutron star usually sits spinning at the center.

The most important observation of SN1987A (for our purposes, at least) was that a pulse of neutrinos was detected, arriving the same night that the supernova was first observed.[40] This is the clearest possible demonstration that a Type II supernova begins with the conversion of ordinary matter into neutron-rich matter. The part of the story involving gravitational collapse thus seems confirmed as well.

Nevertheless, it has proven difficult to reliably estimate the strength of the gravitational waves emitted in this process. The problem is that the physics of the supernova event is extremely complicated, so that it is not easy to make firm predictions about the relevant quantities such as the size of the quadrupole moment and its time derivatives. The most detailed numerical calculations of the course of a supernova event even explicitly make the assumption of spherical symmetry for economy of computation.[39] Saenz and Shapiro[41] pursued a different approach, studying the collapse of a homogeneous ellipsoid, endowed by hand with the physical properties to agree with the detailed spherical models.

The key question is, of course, to what extent is there a departure from a spherical shape that can give rise to large values for $\ddot{I}$? If the departure from spherical symmetry is due to rotational flattening, then the output of gravitational radiation depends on the angular momentum of the collapsing core. For the energy released in the first bounce, Saenz and Shapiro found $\Delta E \propto J^4$ with a peak ΔE of about $10^{-4} M_\odot c^2$ at $J = 3 \times 10^{48}$ erg-sec.

The Crab Pulsar is especially interesting in this context. For years it was the most rapidly spinning neutron star known, with a rotational frequency of about 30.3 Hz. It is found in the middle of the Crab Nebula, M1, a remnant of the great supernova recorded by Chinese astronomers in July 1054. The Crab Pulsar has $J = 2 \times 10^{47}$ erg-sec, although its angular momentum could have been much higher at birth. Even if not, Saenz and Shapiro found that a core with a similar value of angular momentum would grow in eccentricity during succeeding bounces. Many bounces would occur, so that when all bounces are included the gravitational wave output has been substantially enhanced. A core born with the Crab's present angular momentum would radiate more than 3×10^{-6} of its rest mass away gravitationally. This corresponds to peak value of $h = 10^{-23}$ at the distance of the Virgo Cluster. It is important to note, though, that there is no unanimity among the experts that such radiation-enhancing bounces actually occur.

Collapsing cores with very high angular momentum would likely be substantially stronger gravitational wave sources, but there are two important uncertainties to this prediction. One is a physical uncertainty about the history of the collapse, including the competing roles of centrifugal resistance to rapid collapse and of an instability of the core to a bar-shaped distortion that might strongly enhance the strength of the waves. The other great uncertainty is in the probability that a collapsing core might have such a large angular momentum. Some astronomical evidence argues for efficient transfer of angular momentum from the core of a pre-supernova star to its envelope. At first, the discovery of the "millisecond pulsars", starting with PSR1937+214 at $f_{rot} = 642$ Hz, seemed to indicate that pulsars could be born rotating very rapidly. Now, though, the consensus among astronomers is that millisecond pulsars are old objects that have been spun up by mass transfer from companion stars.

The supernova event rate in a galaxy comparable to the Milky Way is believed to be around 3 to 4×10^{-2} yr^{-1}, approximately equal to the estimated pulsar birthrate. This agreement is somewhat surprising since Type II supernovae make up only half of the supernovae, and it is believed that gravitational collapse is not a result of Type I events. Perhaps the discrepancy can be reconciled if pulsars can also be made by the optically "quiet" (but gravitationally "loud"?) collapse of "bare cores" which don't produce a bright flash of light. Mass transfer onto white dwarfs in binaries might be a source of such events.[37] A heliocentric sphere which includes the Virgo Cluster must be searched to find more than one event per year. For efficiencies in the range 10^{-2} to 10^{-4}, this yields $h \approx 10^{-21}$ to 10^{-22}.

3.4.3 *Pulsars*

Even before Supernova 1987A, there was good reason to believe in the gravitational collapse model of Type II supernovae. One of the strongest pieces of evidence was the observation of *pulsars* in the centers of the gas clouds left behind by the supernovae of the past. Pulsars are radio sources that pulse regularly, hence the name, with periods ranging from a small fraction of a second to many seconds. When they were first discovered in 1968 by Jocelyn Bell and Anthony Hewish,[42] the clock-like ticking of the radio signals was so astonishing that the discoverers suspected that they were observing the radio beacons of some extraterrestrial civilization. They even dubbed the radio sources LGM's, short for "little green men". Soon, however, it was realized that a rotating object, emitting radio waves beamed like the light from a lighthouse, could account for the stability of the pulse period.[43] Neutron stars, till then only hypothetical objects, seemed like the best candidates. The history of this exciting episode is nicely told in F.G. Smith's *Pulsars*.[44]

These pulsars have themselves been considered as possible sources of gravitational waves. The signal would consist of a periodic waveform due to the rotation of the presumably non-axisymmetric neutron star. So this is another possible way in which neutron stars may emit gravitational waves. Here, though, the signature of the signal is radically different from that emitted in a binary coalescence or in a supernova.

A simple model of a pulsar is a $1.4\,M_\odot$ neutron star of radius 10 km and moment of inertia $I = 10^{45}$ gm-cm^2, rotating with angular frequency f. If it is to be a source of gravitational waves it must be at least slightly deformed from axisymmetry. Then as it rotates the components of its quadrupole moment would vary; the strength of the gravitational waves thus emitted can be calculated using Eq. 3.6 above. The gravitational radiation emitted by the pulsar should thus be of order

$$h \sim \frac{4\pi^2 G}{Rc^4}\epsilon I f^2,$$ (3.19)

where ϵ is the equatorial ellipticity.

It is this last parameter that is poorly known. A reasonable minimum value is the distortion due to the dipole magnetic field. This can be estimated using the relation[45]

$$\epsilon \approx \frac{U_{mag}}{U_{grav}} \approx \frac{B^2 R^4}{GM^2} \approx 10^{-12}, \tag{3.20}$$

if $B \approx 10^{12}$ gauss, typical of most pulsars. Then the wave amplitude is roughly $h \approx 3 \times 10^{-31} \ (f/1 \text{ kHz})^2 \ (10 \text{ kpc}/R)$. If pulsars are born rapidly rotating then there should be several of the most recent pulsars with such an amplitude present in the galaxy at any time. This rapid rotation question is exactly the same one involved in the argument about the strength of the gravitational wave burst from a supernova. If slow rotation is the norm the signals will be substantially weaker, due to the f^2 dependence of h.

Note that the fastest known pulsar, PSR1937+214, apparently has a low magnetic field, $B \approx 10^8$ gauss, as determined by its period derivative. This is typical of the class of millisecond pulsars. If its distortion is due to magnetic stresses, its gravitational luminosity is probably nearer 10^{14} erg-sec^{-1}, for a value of $h \approx 10^{-39}$. As we mentioned above, it has been proposed that the rapid rotation of this pulsar is due to its having been spun up as it consumed a low mass companion star.[46]

3.4.4 *"Wagoner stars"*

Wagoner[47] described a mechanism for gravitational radiation from a low magnetic field neutron star, such as PSR1937+214, which is in the process of being spun up by accretion from a companion star. As the neutron star acquires angular momentum it reaches a point where it may become unstable to the emission of gravitational radiation. This instability was originally discovered by Chandrasekhar,[48] then studied in detail by Friedman and Schutz.[49] In a rapidly rotating object, gravitational radiation reaction forces tend to drive a non-axisymmetric distortion which in turn enhances the gravitational radiation. This instability is damped by viscosity in the neutron star, so that a steady-state is attained. The resulting amplitude of $\dddot{I}$ thus depends on the poorly-known value of the viscosity of nuclear matter under the conditions of a neutron star. Wagoner gave an estimate assuming one probable value for this viscosity, and noted that an $m = 3$ or 4 distortion (rather than the $m = 2$ quadrupole) is most likely to be excited. The gravitational radiation would then be emitted at $f = 500 \pm 300$ Hz, with an amplitude given

$$h = 3 \times 10^{-27} \left(\frac{1 \text{ kpc}}{D}\right) \left(\frac{1 \text{ kHz}}{mf}\right)^{1/2} \left(\frac{\mathcal{L}_\gamma}{10^{-8} \text{ erg/cm}^2/\text{sec}}\right)^{1/2}, \tag{3.21}$$

where $\mathcal{L}_\gamma$ is the X-ray flux, an indicator of the accretion rate.

The fiducial X-ray flux in Eq. 3.21 is characteristic of strong galactic X-ray sources, such as Sco X-1, which are powered by accretion onto a compact object. A number of such objects are known, and could be targetted for a careful search. Although recent estimates of the viscosity of neutron matter[50] suggest the instability may not result in such large signals, the uncertainty in the prediction comes at an interesting intersection of general relativity with nuclear astrophysics, making this an important mechanism to study.

Note that the amplitude predicted for this source is not necessarily any more out of reach than a millisecond burst of $h \sim 10^{-22}$. Steady monochromatic waves can be seen at much smaller amplitudes, using a combination of long averaging times and resonant enhancement of detector sensitivity. We will begin a discussion of this point in Chapter 4, and return to it in more detail in Chapter 14.

3.4.5 *Black holes*

The existence of black holes is one of general relativity's most dramatic unconfirmed predictions. Even on the basis of Newtonian physics, the eighteenth century physicists Michell and Laplace[51] independently noted that a star of sufficient compactness might have an escape velocity greater than the speed of light. General relativity confirms and enriches this idea. The strangest new feature is that any such object must necessarily collapse under its own weight to infinitesimal size, and thus infinite density. Testing such a prediction is one of the most exciting challenges facing relativistic astrophysics today. We will argue in Chapter 15 that observations of gravitational waves could provide the decisive evidence for the existence of black holes, and one of the best ways to study their properties.

Black holes might form in a gravitational collapse, in a manner similar to the formation of a neutron star discussed above. Just as there is a maximum mass that can be supported by stars or cores of white dwarf form, there is also a maximum mass that can be supported by matter in the form of a neutron star. The exact value of this maximum mass is less well determined, but is believed to be no more than of above about 5 $M_\odot$. Various plausible scenarios could lead to the accumulation of such an amount of material. Thus, it would be surprising if black holes, or something like them, didn't exist.

The birth of a black hole by gravitational collapse, or the collision of two black holes already formed, could be a strong source of gravitational waves. The difficulty with predicting signals from this kind of source is that there is no equivalent of PSR1913+16 to serve as a secure prototype. There are several known objects, Cygnus X-1 and LMC X-3 among them, that are considered likely to be black holes, because they seem too heavy, around 10 $M_\odot$, to be anything else so dark.[52]

Less secure arguments are advanced for the existence of a $10^9 M_\odot$ black hole in the center of the giant elliptical galaxy M87.[53] Indeed, the violent motions of ionized gas clouds in the center of our Galaxy are also believed by several workers to be due to a black hole, in this case of 10^6 $M_\odot$.[54] From the theorist's point of view, massive black holes would also provide a natural explanation for many of the properties of active galactic nuclei.[55] Rees[56] considered the possible evolutionary paths for the dense central regions of galaxies, and found many ways to form black holes, only a few ways to avoid forming them.

Although the case for the existence of black holes has its weaknesses, both theoretical and observational evidence argues for them. The "smoking gun" is likely to come from gravitational wave observations — as we will see below (see Chapter 15), gravitational wave emission from a black hole is inextricably linked to its essential nature as a mass so compressed that it is enshrouded in a horizon from which no signals can escape. No other proposed means of observation can so cleanly study those properties that make a black hole a black hole.

The collision of a pair of black holes emits gravitational waves in a similar manner to a collision of two neutron stars. Detweiler and Szedenits[57] used perturbation theory to study the gravitational radiation from a test particle falling into a Schwarzschild black hole. They found that the amount of rest energy converted into gravitational waves is roughly $\Delta E = A M_1^2 / M_2$, where M_1 is the mass of the test particle, M_2 is the mass of the black hole, and A is a constant of proportionality that depends on the orbital angular momentum. For $J = 0$, A is of order 10^{-2}, while for $J > M_1 M_2 G / c$, A can exceed 10^{-1}. Smarr[58] found that a similar formula applies in the limit that $M_1 = M_2$. If the efficiency $\Delta E / M c^2$ is around 0.1, then the amplitude of the signal is $h \approx 2 \times 10^{-20}$ (1 Mpc/R) $\times (M/M_\odot)$. Details of the gravitational waveform from a collision with realistic parameters still await the solution of many technical problems in numerical relativity. The black hole collision problem was recently designated a Grand Challenge calculation in computational physics, and given a substantial infusion of funds by the U.S. National Science Foundation.

The remaining fact needed to make a prediction of signal strengths is the density of black holes in the Universe. It is here that our knowledge of astrophysics fails us. It is difficult to make confident predictions, yet the opportunities for a far-reaching discovery are nowhere greater than in the case of black holes.

3.4.6 Stochastic backgrounds

One can imagine searching for a stochastic background of gravitational waves in addition to impulsive signals and periodic signals. Operationally, a stochastic signal can be defined as one with roughly constant amplitude (not impulsive) and a broad continuous spectrum (not periodic). Predictions of the possible strength of stochastic gravitational waves are less secure than predictions of the strengths of other kinds of signals. One way to make a stochastic background is to superpose the signals from enough sources so that they overlap, either in time if they are bursts, or in frequency if they are periodic. For the sorts of sources considered previously, the signals become stochastic only at very low values of h.

One interesting possible source is the superposition of the signals from the collapse of a vast cosmic population of black holes. Carr[59] considered several scenarios in which such a population of black holes could form. Depending on the time of formation and the mass distribution, the characteristic frequency of the resulting gravitational waves can vary throughout (or even well beyond) the range 1 Hz to 1 kHz. The energy density allowed in the resulting gravitational wave background may range from 10^{-4} all the way up to 10^{-1} times the closure density of the Universe. The corresponding typical values of h vary from 3×10^{-23} to 10^{-21} near 1 kHz, or 10 times larger if the characteristic frequency is 100 Hz. Thus, the search for a stochastic gravitational background provides a unique means to search for a possible component of the Universe which would be otherwise invisible.

Another source of stochastic gravitational waves could be strong space-time anisotropies arising from quantum gravitational processes in the early Universe. Our understanding of the physics involved is certainly less secure than it is for situations closer to home, but again this is as much an opportunity as a drawback. One interesting idea[60] is that quantum zero-point fluctuations in some modes of the early graviton field may have been strongly amplified during the proposed inflationary stage of the cosmic expansion. Depending on the parameters of inflation, this otherwise unobservable analog of the cosmic microwave background radiation might have been boosted to a detectable level.

3.4.7 *Discussion*

A consideration of astronomical sources of gravitational waves provides the context for the rest of our consideration of gravitational wave detection. In the middle third of this book, we will discuss how it will be possible to construct a detector sensitive enough to register gravitational wave pulses with amplitudes less than or of order 10^{-21}. Toward the end of the book, we will address the astronomical questions again from a different point of view — how to extract information about objects in the sky from the output of gravitational wave detectors.

One final word of caution is necessary: the predictions are based on what we know, or think is plausible. The history of science, and especially astronomy, contains numerous instances where unsuspected phenomena turned out to be dominant. The probability for such an occurrence is highest in cases where a radically new way of looking at the world is first tried. Gravitational wave observations certainly represent such a new look at the sky. Whether or not we will be rewarded with a dramatic surprise is perhaps the most urgent question to be answered by the coming generation of gravitational wave detectors.

Linear Systems, Signals and Noise 4

Gravitational waves, by any practical measure, will cause very weak effects on our measuring devices. Thus, it is natural to make a careful inquiry into the ways in which we will be able to register such small effects, in the presence of a certain unavoidable level of noise. Quantitative discussion of sensitivity is based on the concept of the *signal-to-noise ratio.* An accurate measurement can be made when the effect one is looking for (the *signal*) causes an output from your measuring apparatus that is large compared to the random excursions of the output when no signal is present (the *noise*).

Since so much of the art of gravitational wave detection is concerned with improving the signal-to-noise ratio of measurements, we'll take the time here to review enough of the theory of signal detection to have a working understanding and common language. Discussions of the beautiful theory, much of it invented in the context of the development of radar during World War II, are presented in a number of works, including Davenport and Root's *An Introduction to the Theory of Random Signals and Noise.*[61]

4.1   Characterizing a Time Series

When we search for gravitational waves, we turn on our detector and let it run, recording its output for days, weeks, or maybe months. Thus, an important sort of mathematical object is an input or output as a function of time, called a *time series.*

We can imagine two relevant sorts of functions of time. One is a deterministic function of a predetermined form. In this class might be the gravitational wave signal from a particular kind of source at a particular position and orientation. The second class of function is a random time series that varies from one realization to the next. The noise voltage across a resistor is a member of this latter class, as is the current from a photodiode illuminated by a beam of light.

The reason for drawing this distinction among classes is that it takes different mathematical operations to characterize the regularities of a random time series, than it does to describe a deterministic function of time. Both kinds must be addressed if any sense is to be made of a concept like signal-to-noise ratio, involving a ratio between a deterministic and a random function of time. A good reference for the mathematics of this discussion is Bracewell's *The Fourier Transform and its Applications.*[62]

4.1.1 *The Fourier transform*

Consider first a particular deterministic function of time, $s(t)$. Writing an explicit function for $s(t)$, or tabulating it as a list of values at a closely-spaced set of discrete times, are perfectly good ways of characterizing it. Another way, containing the same information but expressing it in a different and often very suggestive form, is the *Fourier transform* of $s(t)$, called $S(f)$, given by

$$S(f) = \frac{1}{\sqrt{2\pi}} \int_{-\infty}^{\infty} s(t)e^{-i2\pi ft}dt, \tag{4.1}$$

or by a discrete analog of this equation if we are using sampled data. (For the rest of this discussion we'll use "continuous" language, but in the practical world the mathematical objects we'll work with are almost always discretely sampled functions of time.) We'll soon see a number of advantages of thinking in the *frequency domain*, as the world of Fourier transforms is called.

One other note on notation: when doing mathematics, it is often customary to use angular frequencies ω, with units of radians/sec. In the lab, however, one always uses frequencies f denominated in cycles/sec, or Hertz (Hz). Since the mathematical objects described in this chapter are all things one will end up measuring, or using to describe a measurement, I've chosen to write all frequencies here in units of Hz.

4.1.2 *Cross-correlation and autocorrelation*

We will soon want to discuss a way of characterizing a random time series in the frequency domain. But first, it will prove useful to describe two operations we can perform on pairs of functions of time, say $s_1(t)$ and $s_2(t)$. We'll define them here, and point out their utility as we need them. Depending on the application we have in mind, they might be applied to deterministic, random, or a mixed pair of time series.

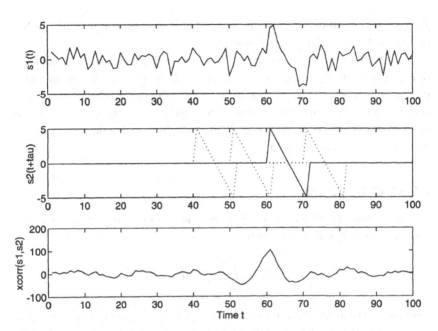

Figure 4.1: The cross-correlation operation. The first time series contains a copy of the template in addition to noise. The second graph shows the template at several time offsets. The last graph shows the cross-correlation as a function of offset.

First, consider the operation called the *cross-correlation* of two functions, defined by

$$s_1 \star s_2(\tau) \equiv \int_{-\infty}^{\infty} s_1(t)s_2(t + \tau)dt. \tag{4.2}$$

The idea behind this definition is explained in Figure 4.1. For each time offset τ, the function $s_2(t)$ is offset in time by τ, multiplied by $s_1(t)$ and the products summed over the entire record. Cross-correlation is a way of measuring the degree of relatedness of the two functions $s_1(t)$ and $s_2(t)$.

When $s_1(t)$ and $s_2(t)$ are the same function, say $s(t)$, then we have the *autocorrelation function*

$$s \star s(\tau) \equiv \int_{-\infty}^{\infty} s(t)s(t + \tau)dt. \tag{4.3}$$

Heuristically, this is a way of measuring the relatedness of a function of time to itself, at various time offsets between the two copies of the function. It clearly will have a maximum at $\tau = 0$, when the function is aligned with itself. Periodic functions will also have maxima at multiples of the period. The width of $s \star s$ is indicative of how rapidly the function changes with time.

Sometimes, such as when our functions of time take on non-zero values for all times, the definitions given above may diverge. In such cases, we will want to replace the integrals from $-\infty$ to ∞ with $\lim_{T \to \infty} (1/2T) \int_{-T}^{T}$. Another normalization convention is also popular for the autocorrelation function; our definition is often replaced by one which is equal to ours divided by $s \star s(0)$, so that the autocorrelation has a value of unity at zero lag τ.

4.1.3 *Convolution*

Another useful and related but distinct operation on two functions of time is the *convolution* of $s_1(t)$ with $s_2(t)$ given by

$$s_1 * s_2(r) \equiv \int_{-\infty}^{\infty} s_1(t) s_2(\tau - t) dt. \tag{4.4}$$

Note the crucial difference in the sign in the argument of s_2, compared to the cross-correlation function. The convolution operator plays a crucial role in describing the behavior of linear systems.

It is possible to carry out all of these mathematical operations on any kind of function, be it deterministic or any particular realization of a random one. But we are particularly interested in the question of what operations most usefully characterize the regularities of a random function of time, that is, which objects have interesting expectation values for an ensemble of random functions. One such useful operation is the autocorrelation function of a random time series. Often, how well correlated the value of a random function is with itself some time later is a useful sort of description. The zero-lag autocorrelation function, $s \star s(\tau = 0)$, is an especially useful number, the mean square value of the function. For the important special case of a time series of zero mean, $s \star s(\tau = 0)$ is identical to the mean square deviation of the function from its mean. In other words, it is a measure of the magnitude of the noise.

4.1.4 *The power spectrum*

While the Fourier transform $S(f)$ is an appropriate way to translate a deterministic time series into the frequency domain, another operation is more appropriate if $s(t)$ is a random time series. In the latter case, it is most useful to calculate the *power spectrum* (also known as the power spectral density) of the time series, defined as the

Fourier transform of the autocorrelation function of the time series, or

$$P_s(f) \equiv \frac{1}{\sqrt{2\pi}} \int_{-\infty}^{\infty} s \star s(\tau) e^{-i2\pi f \tau} d\tau. \tag{4.5}$$

(Here, because we imagine that $s(t)$ is a stationary process defined for all t, we need to use the normalization convention that divides the autocorrelation by the length of the record over which it is computed.) This transform contains all of the information that is coded in the autocorrelation function, expressed in another way. $P_s(f)$ is a measure of the amount of time variation in the time series that occurs with frequency f.

Experimentalists often prefer to think in terms of sines and cosines of positive frequencies, instead of complex exponentials of both positive and negative frequencies. So it is customary to define a *single-sided power spectrum* $s^2(f)$ given by

$$s^2(f) \equiv \begin{cases} 2P_s(f), & \text{if } f \geq 0 \\ 0, & \text{otherwise.} \end{cases} \tag{4.6}$$

From now on, when we use the term "power spectrum", we will be referring to the single-sided version. The mean square value of the function is given by

$$\overline{s^2} = \int_0^{\infty} s^2(f) df. \tag{4.7}$$

4.1.5 *The Periodogram*

There is another route to calculating the power spectrum that is instructive to contemplate. Imagine a particular realization of a random time series $s(t)$ of total length T. Compute the Fourier transform $S(f)$ of this time series. Then $|S(f)|^2/T$ is an object called the *periodogram* (of length T). In the limit as $T \to \infty$ the expectation value of the periodogram is the power spectrum of $s(t)$.

The utility of the approximate equivalence between the periodogram and the power spectrum is the subject of much learned dispute in the statistical literature.[61] There is another theorem to the effect that in the same $T \to \infty$ limit, the variance of the periodogram is equal to its expectation value. In other words, it isn't a very good approximation to the power spectrum after all. This being the case, it may come as a surprise to learn that modern *spectrum analyzers,* wonderfully handy laboratory instruments that let one analyze signals in the frequency domain almost as fluently as an oscilloscope does in the time domain, almost invariably calculate power spectra via the periodogram.

What is going on here? The root of the confusion has to do with what one does with a long data set. The theorem in the previous paragraph warns us not to take one large Fourier transform of the whole data set. One is tempted to do just that, because the longer the time series that is transformed, the more finely one can resolve components of neighboring frequencies. (Yet another theorem[63] says that the resolution goes like $\Delta f \sim 1/T$.) Taking longer data sets to form the periodogram corresponds to trying to form estimates of more frequency resolution elements in the power spectrum, but gives only one estimate of each element.

A much wiser strategy is to break up one's long time series into N chunks of length T/N. Calculate the periodogram corresponding to each chunk, then average them together. A few minutes spent at a spectrum analyzer should be enough to convince you that the average converges very nicely to an estimate of the power spectrum of the limited resolution $\Delta f \sim N/T$.

4.1.6 *Interpretation of power spectra*

In addition to its utility in modern laboratory instrumentation, related to the existence of a powerful algorithm for calculating discrete Fourier transform (the *FFT* or *Fast Fourier Transform*[64]), the periodogram is also useful as a means of gaining intuition into the meaning of the power spectrum. The value of the Fourier transform of a function at a frequency f is a measure of the degree to which that function varies like a sinusoid of frequency f. More precisely, it gives the contribution of a sinusoid of frequency f to a sum of sinusoids that equals the function in question.

Recall, though, that the Fourier transform is a complex-valued function; the real and imaginary parts correspond to the contributions of a cosine and a sine, respectively, to the expansion in sinusoids. This phase information is useful for a deterministic function, but of no relevance whatever for a random function. We might expect a random function to have a tendency to vary at frequency f, but there is no reason for it to have any particular phase. Thus for a random function we expect the Fourier transform itself to have an expectation value of zero, since it might at various times vary as $+\sin 2\pi f t$, $-\sin 2\pi f t$, $+\cos 2\pi f t$, or $-\cos 2\pi f t$. Nevertheless, there will be interesting information coded in $|S(f)|^2$, whose expectation value does not vanish.

This intuition is reinforced by considering another way to build a power spectrum analyzing device. (See Figure 4.2.) For each frequency at which one wants to estimate the power spectrum, imagine applying the noisy time series to a filter that passes sinusoids of only a narrow band of frequencies (say 1 Hz) centered on the frequency of interest. Square the output of the filter (the right diode will do the trick), and average the squared signal by applying it to a low-pass filter of sufficiently long time constant. The output of this final filter is an approximate measurement of the power spectrum of the input signal at the center frequency of the bandpass filter.

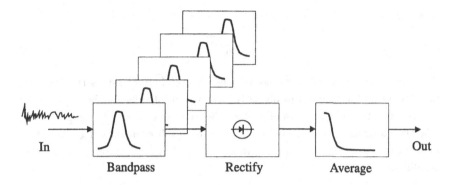

Figure 4.2: A functional block diagram of a filter-based spectrum analyzer.

Before the recent rapid progress in digital electronics, this was the operating principle of commercial spectrum analyzers — they had either a bank of such bandpass filters, or else a single filter whose center frequency was tunable. Today's spectrum analyzers at radio frequencies (too high for practical analog-to-digital conversion) still operate in a similar fashion, using a swept local oscillator to mix various RF frequency components down to the frequency of a fixed bandpass filter, as is done in a heterodyne radio receiver.

Both the periodogram and the bandpass filter method of measuring the power spectrum reinforce our interpretation of the power spectrum $s^2(f)$ as a measure of the extent to which a noisy time series contains a component like a sinusoid (of random phase) of frequency f. If a time series has units of volts, say, then its power spectrum has the units of volts^2-sec or, more suggestively, volts^2/Hz.

These units correspond to another facet of the power spectrum's utility as a mathematical description of a random time series. If one asks how much power (proportional to v^2 in the electrical context, hence the name "power spectrum") passes through a bandpass filter of bandwidth Δf, the answer is given by the product $v^2(f)\Delta f$. That is, in combining noise of neighboring frequency bands, it is the power, or the square of the amplitude, that adds linearly, not the amplitude itself. Thus noise components of different frequencies combine as independent random variables, as they must, as long as the system is linear.

4.1.7 The amplitude spectral density

Experimenters often prefer to work with an object derived from the power spectrum, the *amplitude spectral density* $s(f)$. It is defined simply as the square root of the power spectrum or

$$s(f) \equiv \sqrt{s^2(f)}. \tag{4.8}$$

The amplitude spectral density has units of volts/$\sqrt{\text{Hz}}$ if the time series has units of volts. The advantage of this object is that the unit of volts matches better what we record with a meter or oscilloscope. The disadvantage is the strange-looking "/$\sqrt{\text{Hz}}$", a reminder that taking the square root can't change the fact that independent frequency bands in a noisy time series add in quadrature, not linearly in amplitude.

4.2 Linear Systems

Before completing our discussion of signal detection, there is another mathematical preliminary we need to discuss, the theory of *linear time-invariant systems*. In the present context we will be most concerned with devices called *filters*, but the theory is applicable to a wide range of devices of importance in gravitational wave detectors, including amplifiers, transducers, and feedback loops. Nice references for this material include A. B. Pippard's *Response and Stability*[65] or any good book on control systems, such as Franklin, Powell, and Emami-Naieini's *Feedback Control of Dynamic Systems.*[66]

First, a definition: we are concerned here with devices possessing a single (i.e. scalar) input and a single output, for which there is some linear relationship between the input and output. That is, a linear system is defined by the property that if an input $s_1(t)$ causes an output $v_1(t)$ and $s_2(t)$ causes $v_2(t)$ then an input equal to the sum of s_1 and s_2 gives an output equal to the sum of v_1 and v_2, or

$$s_1(t) + s_2(t) \rightarrow v_1(t) + v_2(t). \tag{4.9}$$

Unless specified otherwise, we will assume that the relationship between the input and output does not change with time, hence the name linear time-invariant systems, or *linear systems* for short.

Consider as an example of a linear time-invariant system the simple one-dimensional damped harmonic oscillator, used as a vibration filter. (See Figure 4.3.) The input x_i is the coordinate of the top of the spring, while the output x_o is the coordinate of the mass.

(This example shows one of the defining features of a *filter* — the input and output are quantities with the same dimensions, in this case meters. The other kind of object with the same feature is the amplifier. The key difference between the two

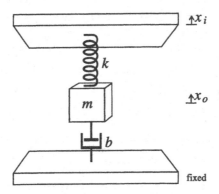

Figure 4.3: A schematic diagram of a mechanical oscillator. Here, we consider the top of the spring to be the input to the system, and the position of the mass to be the output.

is that an amplifier is connected to an external source of energy, in addition to its input, so that more power can be supplied from the output than is supplied at the input. Such *gain* can of course be a function of frequency. The distinction between an amplifier with "shaped" or "tuned" gain and an *active filter* that uses amplifiers as building blocks is often just a matter of degree. In contrast, the term *transducer* is used as a general name for a linear system whose input and output have different physical units.)

The input-output relationship that is the defining feature of a linear system can be specified in a variety of ways. One rather intuitive way is to give the *impulse response* $g(\tau)$, which as the name implies is the output due to an input consisting of a single unit impulse applied at $\tau = 0$. The impulse response $g(\tau)$ can have a wide variety of forms, subject to the limitation that

$$g(\tau) = 0 \quad \text{for } \tau < 0. \tag{4.10}$$

If this limitation were disobeyed, then the system would respond to the input before it was applied — we would say it violated *causality*.

Our example linear system has the equation of motion

$$m\ddot{x}_o + k(x_o - x_i) + b\dot{x}_o = 0. \tag{4.11}$$

Solution of this equation yields an impulse response

$$g(\tau) = e^{-\tau/\tau_0}(f_0^2/f_d)\sin 2\pi f_d \tau, \tag{4.12}$$

where $\tau_0 = 2m/b, f_0 \equiv (1/2\pi)\sqrt{k/m}$, and $f_d \equiv (1/2\pi)\sqrt{(k/m)(1 - b^2/4km)}$. This function is graphed in Figure 4.4. It shows the expected free decay of the system from the initial condition due to the applied impulse.

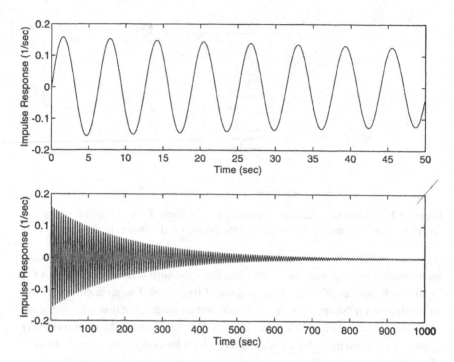

Figure 4.4: The impulse response of our mechanical oscillator. It has a resonant frequency of 0.16 Hz, and an amplitude damping time of 200 sec.

Sometimes, it may be easier to solve the equation of motion for the *step response* $H(\tau)$, the output of the system when a unit step function is applied to the input. In such cases, it is handy to remember the relation

$$g(\tau) = \frac{d}{d\tau}H(\tau). \tag{4.13}$$

The function $H(\tau)$ has the dimensions of the ratio of the output to the input, so for a filter the step response $H(\tau)$ is dimensionless. The impulse response $g(\tau)$ has those dimension divided by one power factor of time. Thus, a filter's impulse response has the units $\sec^{-1}$.

The great utility of the impulse response is that it allows us to write down a simple expression for the response of a linear system to an arbitrary input. This makes sense because we can always imagine representing an arbitrary input as a succession of closely spaced impulses (a "comb") of the proper heights. The linear superposition property, Eq. 4.9, guarantees that the output to such a succession of impulses is just the superposed succession of outputs to the individual impulsive

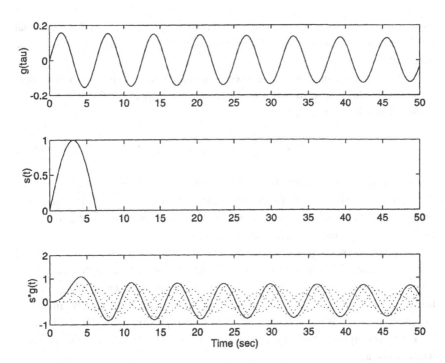

Figure 4.5: The convolution operation. The upper graph shows the impulse response of a linear system. The middle graph represents the time history of the input to the system. In the lower graph, the system's output due to this input is shown.

inputs. This means

$$v(t) = \int_{-\infty}^{t} s(\tau)g(t - \tau)d\tau, \qquad (4.14)$$

or in words, the output of a linear system is the convolution of the input with the impulse response of the system. A graphical representation of this relationship is expressed in Figure 4.5. In the third panel, the dotted curves show the impulse responses to different parts of the input, summing to the solid curve.

Note the sign of the argument of g is that of the convolution operation, not the cross-correlation. To see that this makes sense, consider for example the output at $t = 0$. Then we have

$$v(0) = \int_{-\infty}^{0} s(\tau)g(-\tau)d\tau. \qquad (4.15)$$

The output consists of the sum of the products $s(0) \times g(0)$, $s(-1 \text{ sec}) \times g(1 \text{ sec})$ and so on. Earlier inputs contribute in proportion to the value of the impulse response appropriate to the lapse of time since those inputs were applied. And the fact that $g(\tau)$ is zero for $\tau < 0$ guarantees that inputs in the future do not contribute to the output at the present.

An alternative, and also very powerful, way to represent the action of a linear system is by the Fourier transform of the impulse response, called the *frequency response* $G(f)$. It is defined as

$$G(f) \equiv \frac{1}{\sqrt{2\pi}} \int_{-\infty}^{\infty} g(\tau)e^{-i2\pi f\tau} d\tau. \tag{4.16}$$

$G(f)$ is a complex-valued function of the real frequency f; its real part represents the response "in phase" with a sinusoidal input of frequency f, while the imaginary part gives the "quadrature" component. Often it is represented in polar form, as a magnitude and a phase angle, each a real function of f.

Experimenters often refer to this object by the name *transfer function*, and I will follow that usage in the rest of this book, although purists reserve this latter term for a related object, the *Laplace transform* of the impulse response. The Laplace transform is a cousin of the Fourier transform, in which the purely imaginary exponent in $e^{-i2\pi f\tau}$ is replaced by $e^{-s\tau}$, with s a complex frequency. It gives a more complete description of the dynamic response of a system, at the price of being harder to measure.

The interpretation of the frequency response is suggested by a way of measuring it. Imagine driving (applying an input signal to) the linear system with a unit sinusoid of frequency f, $e^{i2\pi ft}$. The output will also be a sinusoid, $G(f)e^{i2\pi ft}$. Measurement of the magnitude ratio and relative phase angle between the input and output as a function of frequency f is a straightforward procedure for determining $G(f)$ and thus for characterizing a linear system. By contrast, the transfer function defined by the Laplace transform would be determined by measuring the impulse response of the system, then calculating its Laplace transform. The difference lies in the treatment of the transient response. In the frequency response, all transients are assumed to have damped away before measurement of the response to the sinusoidal test signal. By contrast, all of the action for the transfer function is in the transient response.

This interpretation of $G(f)$ can be justified mathematically by considering the input-output relation Eq. 4.14. If we want to know the Fourier transform of the output, $V(f)$, then we need to know the Fourier transform of the convolution integral on the right hand side. The *convolution* theorem[62] states that

$$V(f) = \frac{1}{\sqrt{2\pi}} \int_{-\infty}^{\infty} e^{i2\pi ft} dt \int_{-\infty}^{t} s(\tau)g(t-\tau)d\tau = S(f)G(f). \tag{4.17}$$

Thus we have

$$G(f) = \frac{V(f)}{S(f)}, \tag{4.18}$$

showing that the frequency response is the complex ratio of the system's output to its input. Equation 4.18 represents the way one might measure $G(f)$ by hand, applying inputs S as sinusoids of various frequencies f and measuring the output magnitude ratio and phase difference with respect to the input. A modern FFT spectrum analyzer carries out an equivalent procedure. Exciting the system with noise of many frequencies the analyzer calculates the Fourier transforms of the input and output, then forms the ratio $V(f)/S(f)$ to determine the frequency response $G(f)$.

Comparison of Eqs. 4.17 and 4.18 shows the powerful advantage of working in the frequency domain — in Fourier space the output of a linear system is merely the product of the input and the frequency response. No convolution integral need be evaluated.

4.2.1 *Bode plots*

It is customary to display the frequency response in a form called a *Bode plot*, after H. W. Bode of Bell Labs. In a Bode plot, the logarithm of the magnitude of the frequency response is graphed against frequency f, with f also plotted on a logarithmic scale.

Below this graph, the phase angle is also plotted, in a linear scale usually running between $180°$ and $-180°$ or $200°$ and $-200°$, against the same logarithmic frequency axis. (Recall that any angle can be represented in the interval $-180° \le \phi < 180°$.) The magnitude scale is traditionally scaled in decibels (or dB) defined by

$$\text{Mag(dB)} = 20 \log_{10} |G(f)|. \tag{4.19}$$

4.2.2 *Frequency response example*

Let's compute the frequency response $G(f)$ of the model linear system of Figure 4.3. We could find it by direct Fourier transformation of the impulse response $g(\tau)$. There is another method of finding it that is such a useful trick that it is worth demonstrating here. Starting with the equation of motion Eq. 4.11, we assume an input of the form

$$x_i(t) = x_i(f)e^{i2\pi f t}. \tag{4.20}$$

Then, because the system is described by a linear differential equation, we know that the output x_o will also be of sinusoidal form. Next, recall that the first time

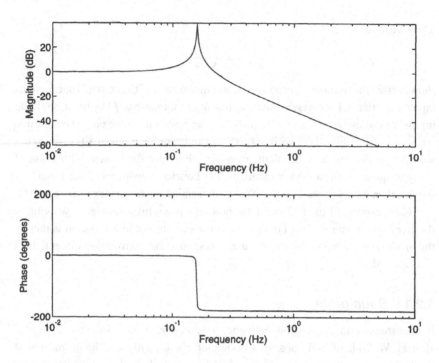

Figure 4.6: A Bode plot of the transfer function of the mechanical oscillator.

derivative of $e^{i2\pi ft}$ is $i2\pi f e^{i2\pi ft}$, and the second derivative is $-(2\pi f)^2 e^{i2\pi ft}$. Now, substitute Eq. 4.20 into the equation of motion, and find

$$-m(2\pi f)^2 x_o + k(x_o - x_i) + i2\pi f b x_o = 0. \tag{4.21}$$

Solution of the equation for $G(f) = x_o/x_i$ is now simply a matter of algebra. We obtain

$$G(f) = \frac{k}{k + i2\pi fb - m(2\pi f)^2}. \tag{4.22}$$

A Bode plot is shown as Figure 4.6. $G(f)$ contains the same information as $g(r)$, but in a form especially easy to obtain, either by measurement or by solution of the equation of motion.

4.3 The Signal-to-Noise Ratio

We have now laid the groundwork for an elementary discussion of the detection of signals in the presence of noise. Our discussion will focus on the concept of the *signal-to-noise ratio*, or *SNR*. This a dimensionless figure of merit for a measurement. In Chapter 14 we will discuss the interpretation of the signal-to-noise ratio in terms of the probability of a false inference that a signal has been detected. For now, it will suffice to remark that a measurement characterized by a signal-to-noise ratio of order one is not of much use, while signal-to-noise ratios of 10 or greater represent detections of some confidence.

But how can we even construct a dimensionless ratio of signal strength to noise amplitude? It may seem difficult, since they are measured in different units. As we saw above, signals are measured in volts, for example, or perhaps volts/Hz in Fourier space. On the other hand noise, as a stochastic process, is characterized by a power spectrum, with units of volt^2/Hz (or an amplitude spectral density in $\text{volt}/\sqrt{\text{Hz}}$.)

The answer lies in a more careful consideration of what one does when one makes a measurement. As in the sections above, we'll discuss this question first in the time domain, then give the insights available from frequency-domain concepts. For now, we'll concentrate on how to look through a single output time series for a signal of short duration. We'll return to some of these questions in more detail in Chapter 14; there we'll consider coincidence techniques, searches for steady periodic signals, and other topics.

4.3.1 *Noise statistics*

Think of the output $v(t)$ of some noisy detector, for example a noisy trace on a chart recorder. Better yet, imagine that such a record has been digitized; then the record consists of a string of numbers, representing the detector output at a set of closely spaced times. When no signal is present, the output fluctuates randomly with a known power spectrum. We assume the noise has a mean value of zero.

In order to distinguish a signal from the noise, we have to assume that we would know the signal when we saw it. Typically, our description of the signal is in the form of a particular function of time, $s(t)$. (Usually, $s(t)$ can only be predicted up to some multiplicative constant α.) In this context, we call the function $s(t)$ the *template*. Signal detection is the process of looking through a noisy record for a pattern resembling the template, occurring at a strength unlikely to be due to noise alone.

With a digitized record, it is simple to conceive of a procedure to search for a match between the template and the time record. We form the cross-correlation integral between the template and the time record, and evaluate it for every possible distinct time at which the signal could have arrived. That is, we compute

$$v \star s(t) = \int_{-\infty}^{\infty} v(\tau)s(t + \tau)d\tau. \qquad (4.23)$$

First, consider a noise-free time series. Clearly, when the template $s(t)$ coincides with an occurrence of a signal $v(t) = \alpha s(t - t_0)$ in the time record, the cross-correlation will have a value of $\int_{-\infty}^{\infty} \alpha s^2(\tau)d\tau$, greater than if it were offset from perfect coincidence. If the template were aligned with a section in which the output were zero (no signal and no noise) then the cross-correlation would be zero.

Since we are considering a noisy detector (they all are, at some level), the cross-correlation of the template with the output is seldom zero. Instead, the function $v \star s(t)$ is another noisy time series. If the noise in the detector output has a *Gaussian probability distribution*

$$P(v) = \frac{1}{\sqrt{2\pi\sigma^2}} e^{-v^2/2\sigma^2}, \tag{4.24}$$

often a good approximation, then the probability distribution of $v \star s$ is also a Gaussian, as long as no signals are present.[67] Imagine forming a histogram of values of $v\star s$. The vast bulk of values will fall in the Gaussian distribution. (See Figure 4.7.)

Sometimes, if we are lucky enough to have a signal of sufficient size present, the detector output will match the template well. Then $v \star s$ will have an especially

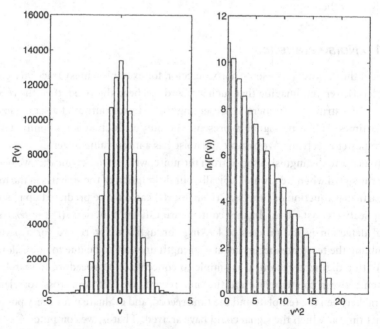

Figure 4.7: Two versions of the histogram of a variable with a Gaussian probability distribution. At left it is plotted with linear scales on both axes; on the right, the scales are chosen so that a Gaussian exhibits a straight line.

large value. On the histogram, it will fall well outside the bins filled with noise events. Any such "outlier" is a putative signal, since if the cross-correlation is large it is unlikely to have been caused by noise alone. Whether any particular such event actually corresponds to a signal of the sort one is looking for, or alternatively whether it corresponds to some sort of competing effect that we might call *interference*, is a matter statistical analysis cannot answer by itself.

We are finally ready to define the signal-to-noise ratio. For any given template, we can characterize the noise by

$$N^2 \equiv \sqrt{\langle (v \star s(\tau))^2 \rangle}, \qquad (4.25)$$

the root mean square value of the cross-correlation between the noise and the template. (The triangular brackets $\langle \rangle$ indicate averaging over time.) This is a measure of the width of its distribution in a histogram. We can characterize the strength S^2 of the signal present at any time t by the cross-correlation between the expected form of the output $s(\tau)$ and the output $v(t)$, that is by

$$S^2 \equiv |v \star s(t)|. \qquad (4.26)$$

The signal-to-noise ratio, or SNR, is the square root of the ratio of this measure of the amount of signal present to the expected value due to noise alone, or

$$\text{SNR} \equiv \sqrt{S^2/N^2}. \qquad (4.27)$$

Thus we have found a sensible way of normalizing the signal to the noise.

The SNR is a monotonic measure of how improbable it is to see a given value of the cross-correlation, under the assumption that the output contained just noise, with no signal. Thus, a large SNR is indicative that something other than noise is present in the time series.

4.3.2 *Matched templates and matched filters*

In the previous section, we implicitly assumed that the best way to pick a signal out of noise is to cross-correlate the noisy detector output with a template that is a true copy of the signal. This makes intuitive sense, since what we are trying to do is pick out something of that particular form, from a din of polymorphous noise. It can be proven[68] that this is actually the optimum choice for a template, as long as we are dealing with *white noise*, that is noise whose spectral density is independent of frequency. Then, we dignify this procedure with the term *matched template*. The situation is a little more complicated when the noise isn't white, as we'll see, but not too much.

Before we handle that case, it is interesting to describe in more detail how one might implement the detection procedure. In so doing, we'll make more connections with the mathematical language we discussed earlier. One way to implement the procedure is to carry it out with a computer pretty much as it has been described. The cross-correlation function is straightforward to compute. It is especially well-suited to the common situation in which you'd like to record the detector output now, and figure out later what sort of signals to look for.

But it hasn't always been so simple to digitize and record a long time series of data. And, there are also situations when you'd like to know as soon as possible (in *real time*) whether a signal was present. This is a description of the situation in the classic radar problem. The development and implementation of radar during World War II was the context in which the theory of signal detection was first thought out carefully and it still represents a prototypical problem. The solution worked out in the 1940's is still often the most practical one; it is also instructive.

How can one construct an analog real-time device that carries out the cross-correlation operation? The answer seems obvious once we recall the close relation between the cross-correlation and the convolution integrals. A linear system, in effect, carries out the convolution integral between its input and its impulse response. The only difference is that the sign of the argument of the impulse response in the convolution is the opposite of the sign of the analogous template in the cross-correlation.

This complication can be easily resolved, and the analogy made complete, if we create a linear system, called a *matched filter*, whose impulse response is the time reverse of the signal $s(\tau)$ that we are looking for. If we apply the noisy output of our detector to the input port of our matched filter, then scanning the filter's output for large values is exactly equivalent to looking for unexpectedly large values of the cross-correlation with a matched template.

As before, we can describe this same process in the frequency domain. The matched filter's impulse response $g(\tau) = s(-\tau)$ has the Fourier transform

$$G(f) = e^{-i2\pi f t_0} S^*(f), \qquad (4.28)$$

where the time t_0 is a parameter representing the duration of the time series used by the filter for forming the cross-correlation. The properly filtered noisy output thus has the form $x(f) \propto G(f)V(f)$. This transformation to the frequency domain leads to a qualitatively different kind of insight. The matched filter's proportionality to $S^*(f)$ shows that it is a filter that passes frequencies in which the signal is rich much more readily than those frequencies in which it is deficient.

Now we are in a position to discuss how to create the equivalent of a matched filter in the more usual case when the noise isn't white. The answer is based on an idea carried over from the white noise case, that one should emphasize frequencies at which the ratio of signal amplitude to noise power is large, Hertz by Hertz, to the exclusion of frequencies at which this "SNR density" is small. The optimum filter has a transfer function[69]

$$G(f) \propto \frac{e^{-i2\pi f t_0} S^*(f)}{N^2(f)}. \tag{4.29}$$

One way to see heuristically that this is a sensible result is to think of the function of this filter in two successive steps: first, a filter corresponding to the denominator that "whitens" the noise, then a filter corresponding to the numerator, that applies the prescription of the white noise case, Eq. 4.28.[70]

It is, of course, one thing to write tke transfer function of the desired filter, and another thing to devise an analog filter circuit with that transfer function. The problems come in two flavors, at least. One is to be able to invent the right structure, assuming one does exist — this is the *synthesis problem*. The second is the question of whether in fact a linear system is allowed by causality to have the required transfer function: this is called the question of *realizability*. A filter is said to be unrealizable if it requires information about future values of its input to generate its output. Typically, such problems can be avoided, at least largely, by choosing to design a filter that will give its output some time after the signal has passed — that way it gets to "see" the whole signal waveform without violating causality.

Both of these problems tend to disappear when we look for signals by non-real time digital cross-correlation with an optimum template. The inverse Fourier transform of $G(f)$ gives the template. If that template covers a large interval of the time record, so be it.

4.3.3 *SNR rules of thumb*

In a field like gravitational wave astronomy, we may not have a very good idea of the exact signal waveform arriving at our detector, in spite of the best efforts of our theorist colleagues. So it is worthwhile to discuss what can be achieved with rough-and-ready approximations to the optimum. This will also lead us to a rule of thumb for estimating up to about a factor of 2 the signal-to-noise ratio for a number of archetypal classes of measurements.

A frequency domain picture works most simply here. Even when we don't know the exact waveform, we usually have a qualitative to semi-quantitative idea of which frequencies are well represented in the signal. What is important is to be able to characterize the signal in terms of a characteristic frequency and a bandwidth.

Here is a "poor man's" table of Fourier transform pairs: A single full cycle of oscillation has a characteristic frequency f_c of order the inverse of the duration of the signal, and a bandwidth Δf of the same order as the characteristic frequency. A signal burst consisting of N cycles of characteristic frequency f_c has a bandwidth Δf of order f_c/N. This can be thought of as a classical uncertainty relation for waves.

Armed with this information, we could always create a crudely optimum filter, designed to pass the signal's characteristic band of frequencies, and to reject noise outside that band. If we know that some of those signal-rich frequencies contain excess noise power, then we would narrow the filter to pass only the lower noise parts of the signal band.

When we use such a filter for searching for the appropriate signal, what will be the signal-to-noise ratio? Characterize it by an amplitude s_0 typical of its stronger parts. The mean square noise amplitude through the filter will be, approximately, the average value of the noise power spectrum in the filter band, multiplied by the filter bandwidth. This means the signal-to-noise ratio will be about

$$SNR \sim \frac{s_0}{\sqrt{n^2(f)\Delta f}}. \tag{4.30}$$

This is one of the most basic rules of thumb of experimental physics.

One thing this rule immediately makes clear is the benefit of being able to average a signal for a long time. If a signal is only present for a short time, say a 1 millisecond long burst from a supernova core collapse, the effective noise bandwidth is wide, 1 kHz in this case. All of the noise power in that wide band will contribute to the rms noise that competes with our ability to recognize a signal. The numerical value of the rms noise is $\sqrt{\Delta f}$ times the value of the noise amplitude spectral density in, say, volts/$\sqrt{Hz}$. Looking for a signal 1 millisecond long in noise of 1 $\mu V/\sqrt{Hz}$ means competing with an rms noise of about 30 μV.

On the other hand, if we were looking for a signal that oscillated at some particular frequency for a duration of, say, 100 sec, then we can expect to detect a signal of much smaller amplitude. The rms noise passed by the matched filter for the long duration signal will be about 1 $\mu V/\sqrt{Hz}$ divided by the square root of 100 sec, or about 0.1 μV. The advantage of being able to average is striking.

4.3.4 The characteristic amplitude

When we compare the strengths of two different signals, one way is to just compare their peak amplitudes. But if we want to compare how easily the signals could be detected, it makes more sense to include in the comparison the benefit of averaging. All else being equal, the benefit can be measured by counting the number of cycles

N spent by the signal near to its maximum amplitude. Then the more useful way to compare the strengths of two signals is by looking at the magnitude not of s_0 but of the signal's *characteristic amplitude* $s_c \equiv s_0 \sqrt{N}$. For a more formal definition of an h_c specialized for gravitational wave amplitudes, see Thorne's essay in *300 Years of Gravitation.*[9]

Optical Readout Noise $\qquad$ 5

Now we are ready to confront the question, "How well can an interferometer work?" The precision likely to be required is daunting. Metric perturbations h (and thus fractional changes in light travel time) of the order of 10^{-21} or smaller are typical of predictions of wave strength. It is convenient to express the wave's effect as an equivalent motion of the test masses. Recalling the "proper reference frame" description, we have

$$\Delta x = \frac{1}{2}Lh, \tag{5.1}$$

or

$$\Delta\phi = \frac{4\pi}{\lambda}Lh. \tag{5.2}$$

In an interferometer the size of Michelson and Morley's, with total length $L_{opt}/\lambda = 4 \times 10^7$ ($\lambda \approx 0.5~\mu$m), $h = 10^{-21}$ would cause a shift of only 2.5×10^{-13} of a fringe. Michelson and Morley only claimed to be able to detect shifts of $\lambda/20$ or so. Even if one could construct an interferometer with $L_{opt} = 1000$ kilometers, or $L_{opt}/\lambda = 2 \times 10^{12}$, $h = 10^{-21}$ only corresponds to an optical path length change of 10^{-8} of a fringe. Is it possible to confidently detect such a tiny change in path length?

In this chapter, we'll see how to answer this question. We'll look at the basic physics issues involved, in the context of a simple interferometer in which the light only makes one round trip in an arm. Then, in Chapter 6, we'll show how to generalize the results to more realistic gravitational wave interferometers, which make use of Michelson and Morley's trick of folding up long light paths into comparatively short arms.

5.1   Photon Shot Noise

A key to finding the answer to the question of sensitivity is to remember that we can (and will) use the variant of the Michelson interferometer in which the recombining wave fronts are made strictly parallel. In this case the pattern of bright and dark "fringes" observed by Michelson and Morley widens out into a single spot over which the phase difference between the two beams is a constant.

This spot will brighten or darken as the phase difference is adjusted. The output power is given by

$$P_{out} = P_{in} \cos^2 (k_x L_x - k_y L_y), \qquad (5.3)$$

a graph of which appears in Figure 5.1. In other words, what we have is a device in which the path length difference between the two arms can be determined (up to an integer number of wavelengths) by a careful measurement of the optical power at the interferometer output.

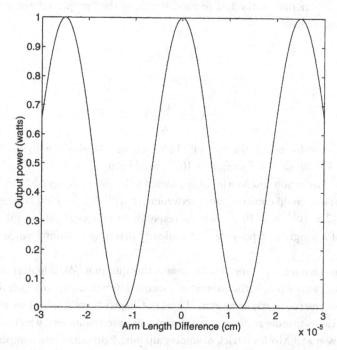

Figure 5.1:   The output power from a Michelson interferometer, as a function of the difference in arm lengths. In this example, the total optical power is 1 watt, and the wavelength is 0.5 micron.

This means we can replace the question "How small a gravitational wave amplitude can we detect?" with the question "How small a change in optical power can we detect?" This latter question suggests that there may in fact be a fundamental limit to the measurement precision.

To see why, recall that light comes in finite sized chunks called photons. Measuring optical power is equivalent to determining the number of photons arriving during a measurement interval. Whenever we count a number of discrete independent events characterized by a mean number $\bar{N}$ per counting interval, the set of outcomes is characterized by a probability distribution $p(N)$ called the *Poisson distribution*,

$$p(N) = \frac{\bar{N}^N e^{-\bar{N}}}{N!}. \tag{5.4}$$

(This is also colloquially referred to as "counting statistics".) When $\bar{N} \gg 1$, the Poisson distribution can be approximated by a Gaussian distribution with a standard deviation σ equal to $\sqrt{\bar{N}}$.

Consider an experiment in which we try to determine the rate of arrival of photons $\bar{n}$ (with units of sec^{-1}), by making a set of measurements each lasting τ seconds. The mean number of photons per measurement interval is $\bar{N} = \bar{n}\tau$. The fractional precision of a single measurement of the photon arrival rate (or, equivalently, of the power) is thus given by

$$\frac{\sigma_{\bar{N}}}{\bar{N}} = \frac{\sqrt{\bar{n}\tau}}{\bar{n}\tau} = \frac{1}{\sqrt{\bar{n}\tau}}. \tag{5.5}$$

This says that if we were to try to estimate $\bar{n}$ from measurements for which $\bar{n}\tau \sim 1$, then the fluctuations from instance to instance will be of order unity. But, if $\bar{n}\tau$ is very large, then the fractional fluctuations are small.

Let's carry through the calculation for the power fluctuations, and thence to the noise in measurements of h. Each photon carries an energy of $\hbar\omega = 2\pi\hbar c/\lambda$. (We choose not to absorb the 2π into Planck's constant so that we can reserve the symbol h for the gravitational wave metric perturbation.) If there is a power P_{out} at the output of the interferometer then the photon flux at the output will be

$$\bar{n} = \frac{\lambda}{2\pi\hbar c} P_{out}. \tag{5.6}$$

Now we need to specify the operating point of the interferometer, or in other words the mean position of the interferometer in the diagram of Figure 5.1. (In order for this concept to be meaningful, the state of the interferometer must be nearly fixed, either because the noise is intrinsically small or because a servomechanism keeps it fixed. Although the latter will always be the case in practice, let's assume for simplicity's sake that the former is true). The most sensible operating point would appear to be $P_{out} = P_{in}/2$. There, midway between maximum power and $P_{out} = 0$, the sensitivity dP_{out}/dL to arm length shifts is a maximum. At that point,

$$\frac{dP_{out}}{dL} = \frac{2\pi}{\lambda}P_{in}. \tag{5.7}$$

We can also consider this to be the sensitivity to the test mass position *difference* δL, since the interferometer is equally sensitive (with opposite signs) to shifts in the length of either arm.

Now consider the fluctuations in the mean output power $P_{out} = P_{in}/2$, averaged over an interval τ. The mean number of photons per measurement is $\bar{N} = (\lambda/4\pi\hbar c)P_{in}\tau$. Thus we expect a fractional photon number fluctuation of $\sigma_{\bar{N}}/\bar{N} = \sqrt{4\pi\hbar c/\lambda P_{in}\tau}$. Since we are using the output power as a monitor of test mass position difference, we would interpret such statistical power fluctuations as equivalent to position difference fluctuations of a magnitude given by the fractional photon number fluctuation divided by the fractional output power change per unit position difference, or

$$\sigma_{\delta L} = \frac{\sigma_N}{N} \Big/ \frac{1}{P_{out}} \frac{dP_{out}}{dL} = \sqrt{\frac{\hbar c\lambda}{4\pi P_{in}\tau}}. \tag{5.8}$$

Recall from Chapter 2 that we can describe the effect of a gravitational wave of amplitude h as equivalent to a fractional length change in one arm of $\Delta L/L = h/2$, along with an equal and opposite change in the orthogonal arm. The net change in test mass position difference is $\delta L = Lh$, so if we interpret brightness fluctuations in terms of the equivalent gravitational wave noise σ_h, we have $\sigma_h = \sigma_{\delta L}/L$, or

$$\sigma_h = \frac{1}{L}\sqrt{\frac{\hbar c\lambda}{4\pi P_{in}\tau}}. \tag{5.9}$$

There is no preferred frequency scale to this noise; the arrival of each photon is independent of the arrival of each of the others. Note also that the error in h scales inversely with the square root of the integration time. These facts can be summarized by rewriting Eq. 5.9 as the statement that the *photon shot noise* in h is described by a white amplitude spectral density of magnitude

$$h_{shot}(f) = \frac{1}{L}\sqrt{\frac{\hbar c\lambda}{2\pi P_{in}}}. \tag{5.10}$$

(It is no accident that the amplitude spectral density looks almost identical to the rms fluctuation with τ set equal to 1 second. The correspondence would be exact if we were using two-sided spectral densities, but we pick up a factor of $\sqrt{2}$ from our preference for one-sided spectra. All of the mathematical details are explained in, for example, Davenport and Root's discussion of the Schottky formula for the shot noise in a vacuum diode.[71])

To set the scale, we can rewrite the rms error in h as

$$\sigma_h = 3.7 \times 10^{-22} \left(\frac{1000 \text{ km}}{L} \right) \sqrt{\frac{\lambda}{0.545 \, \mu\text{m}}} \sqrt{\frac{1 \text{ watt}}{P_{in}}} \sqrt{\frac{10 \text{ msec}}{\tau}}, \qquad (5.11)$$

or the spectral density of h as

$$h(f) = 5.2 \times 10^{-23} \text{ Hz}^{-1/2} \left(\frac{1000 \text{ km}}{L} \right) \sqrt{\frac{\lambda}{0.545 \, \mu\text{m}}} \sqrt{\frac{1 \text{ watt}}{P_{in}}}. \qquad (5.12)$$

It is clear that there is little margin to spare, if we want to be able to confidently detect, and then study, burst signals with amplitudes of 10^{-21} or below. We've already pushed well beyond Michelson and Morley's effective arm length $L = 11$ m. The arm length $L = 1000$ km is nearly the optimum length for a burst of duration 10 msec. The wavelength $\lambda = 0.545 \, \mu$m is that of green argon ion lasers; the power $P_{in} = 1$ watt is a conservative estimate for the power available from such a laser. Laser and mirror technology do not hold out much hope at present of substantial reduction of λ.

So our hopes for advancing beyond this entry-level sensitivity will rest on maximizing the input optical power P_{in}. Fortunately, this does indeed seem possible. Sufficiently stable lasers producing of order 100 Watts are at an advanced stage of development.[72] And, as we will see in Chapter 12, there are clever variations on the standard interferometer we've been discussing so far, whose shot noise is substantially lower than that given by Eq. 5.12. Thus there is every reason to believe that astronomically interesting shot noise levels can be achieved.

5.2 Radiation Pressure Noise

If shot noise were the only limit to precision determined by the optical power P_{in}, then we could in principle achieve arbitrary precision simply by using a sufficiently powerful laser. But in addition to the substantial technical problems of high power lasers, this line of reasoning neglects one of the deep truths of quantum mechanics. We've treated so far the limit that a quantized world sets on the precision of a

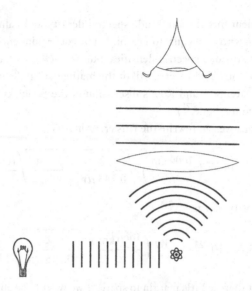

Figure 5.2: A thought-experiment version of the Heisenberg microscope. Light from the lamp illuminates an atom. Reflected light is imaged through the optical system of the microscope.

measurement. We need to *complement* (to use a term of Niels Bohr's) that discussion with a description of how the measurement process disturbs the system under measurement. It is convenient to make a mental division of a gravitational wave interferometer into two parts. Call the nearly freely-falling mirrored test masses (and the space-time between them) the "system to be measured", and the laser, light beams, and photodetector the "measuring apparatus". There is a deep analogy between such an interferometer and the archetypal quantum mechanical measurement problem called the "Heisenberg microscope". (See Figure 5.2.) Bohr gave a particularly clear description of it, using a semi-classical treatment. In his 1928 essay "The Quantum Postulate and the Recent Development of the Quantum Theory", Bohr wrote[73]:

> In using an optical instrument for determinations of position, it is necessary to remember that the formation of the image always requires a convergent beam of light. Denoting by λ the wave-length of the radiation used, and by ϵ the so-called numerical aperture, that is, the sine of half the angle of convergence, the resolving power of a microscope is given by the well-known expression $\lambda/2\epsilon$. Even if the object is illuminated by parallel light, so that the momentum h/λ of the incident light quantum is known both as regards magnitude and direction, the finite value of the aperture will prevent an exact knowledge of the recoil accompanying the

scattering. Also, even if the momentum of the particle were accurately known before the scattering process, our knowledge of the component of momentum parallel to the focal plane after the observation would be affected by an uncertainty amounting to $2\epsilon h/\lambda$. The product of the least inaccuracies with which the positional co-ordinate and the component of momentum in a definite direction can be ascertained is just given by [the uncertainty relation].

We are hardly dealing with a microscopic system here: the test masses are likely to range in size from 10 kg to 1 tonne or so. Yet because we aspire to such extreme precision of measurement, it is crucial to consider the sort of quantum effects usually relevant only for processes on the atomic scale. Note that we are not satisfied to know our test masses' precise positions at one moment only; we want to know the history of the path length difference of the interferometer. Perturbations of the momenta of the masses cannot be ignored, therefore, since the value of the momentum at one time affects the position later.

In the Heisenberg microscope, the phenomenon conjugate to the registration of the arrival of a photon that has bounced off an atom is the recoil of the atom caused by the change in the photon's momentum upon reflection. In a gravitational wave interferometer we register an arrival rate of photons that we interpret as a measurement of the difference in phase between electromagnetic fields returning from the two arms. We can recognize the conjugate phenomenon by looking for a fluctuating recoil that can affect the same degree of freedom that we measure. Fluctuating radiation pressure on the test masses causes them to move in a noisy way. The resulting fluctuation in the length difference between the two arms shows how this effect can alter the phase difference between light arriving from the two arms; this identifies it as the conjugate phenomenon.

To estimate the size of this effect, first recall that the force exerted by an electromagnetic wave of power P reflecting normally from a lossless mirror is

$$F_{rad} = \frac{P}{c}. \tag{5.13}$$

The fluctuation in this force is due to shot noise fluctuation in P, discussed in the previous section. That is

$$\sigma_F = \frac{1}{c}\sigma_P, \tag{5.14}$$

or, in terms of an amplitude spectral density

$$F(f) = \sqrt{\frac{2\pi\hbar P_{in}}{c\lambda}} \tag{5.15}$$

independent of frequency.

This noisy force is applied to each mass in an arm. For now, let us consider a simple "one-bounce" interferometer like the first version of the Michelson interferometer. We will allow the mirrors at the ends of the arms to be free masses, but in this example assume that the beam splitter is much more massive than the other mirrors. The fluctuating radiation pressure from the power $P_{in}/2$ causes each mass to move with a spectrum

$$x(f) = \frac{1}{m(2\pi f)^2} F(f) = \frac{1}{mf^2} \sqrt{\frac{\hbar P_{in}}{8\pi^3 c\lambda}}. \tag{5.16}$$

The power fluctuations in the two arms will be anti-correlated. (In a semi-classical picture, one additional photon into one arm means one less into the other arm.) This doubles the effect on the output of an interferometer, since the phase shift is proportional to the difference in length of the two arms. Thus we have the *radiation pressure noise*

$$h_{rp}(f) = \frac{2}{L} x(f) = \frac{1}{mf^2 L} = \sqrt{\frac{\hbar P_{in}}{2\pi^3 c\lambda}}. \tag{5.17}$$

Thus we have two different sources of noise associated with the quantum nature of light. Note that they have opposite scaling with the light power — shot noise declines as the power grows, but radiation pressure noise grows with power.

If we choose to, we can consider these two noise sources to be two faces of a single noise; call it *optical readout noise*, given by the quadrature sum

$$h_{o.r.o}(f) = \sqrt{h_{shot}^2(f) + h_{rp}^2(f)}. \tag{5.18}$$

At low frequencies, the radiation pressure term (proportional to $1/f^2$) will dominate, while at high frequencies the shot noise (which is independent of frequency, or "white") is more important (See Figure 5.3.) We could improve the high frequency sensitivity by increasing P_{in}, at the expense of increased noise at low frequency. At any given frequency f_0, there is a minimum noise spectral density; clearly, this occurs when the power P_{in} is chosen to have the value P_{opt} that yields $h_{shot}(f_0) = h_{rp}(f_0)$. The power that gives this relation is

$$P_{opt} = \pi c\lambda m f^2. \tag{5.19}$$

P_{opt} is typically quite large. Within this (too simple) interferometer model, let's make an indicative number, Take $m = 10$ kg, $f = 100$ Hz, and $\lambda = 0.545 \ \mu$m. Then, P_{opt} is about 0.5 Megawatt. Even though this number is reduced by the optical path folding schemes we'll talk about in the next section, it will still be true that early

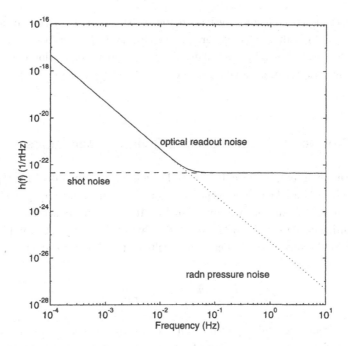

Figure 5.3: The optical readout noise of a Michelson interferometer. Here, each test mass has a mass of 10 kg, and is located 500 km from the very massive beam splitter. The interferometer is illuminated with 5 watts of optical power.

generation interferometers will generally run well below their optimum power, and so will have shot noise larger than the optimum.

When we plug this expression for P_{opt} into our formula for $h_{o.r.o.}$ we find

$$h_{QL}(f) = \frac{1}{\pi f L}\sqrt{\frac{\hbar}{m}}. \tag{5.20}$$

We have renamed this locus of lowest possible noise $h_{QL}(f)$, for "quantum limit", to emphasize its fundamental relationship to quantum mechanical limits to the precision of measurements. Note that the expression does not depend on P_{in} or λ, or any other feature of the readout scheme, even though such details were useful for our derivation. Thus, this examination of the workings of our Heisenberg microscope provides an instrument-specific derivation of Heisenberg uncertainty principle. And it reminds us of the truth Bohr's remarks expressed, that in any measurement the uncertainty principle emerges from the specific mechanism of the measurement.

(A word of warning about notation: the left hand side of Eq. 5.20 is written as if this "quantum limit" noise had a particular spectral density. But note that the object we call $h_{QL}(f)$ is instead the locus of the lowest possible points of the family of spectra $h_{o.r.o.}(f)$, with that family parametrized by, say, f_0. For this reason, $h_{QL}(f)$ is sometimes called a "pseudo-spectral density"[74]; we often want to plot it on a graph of noise spectra, but the actual noise spectra will always be larger, except at one frequency.)

5.3   Shot Noise in Classical and Quantum Mechanics

We determine the output power by letting the light fall on a photodetector, typically a silicon photodiode. An individual photon promotes, with probability η, an electron to the conduction band in the semiconductor. (It is not hard to achieve $\eta \approx 0.8$.) The resulting current of photoelectrons is the physical quantity we can amplify and measure. So the shot noise of interest is actually that of the photocurrent

$$I_p = \eta \bar{n} e = \eta e P_{in} \frac{\lambda}{2\pi \hbar c}. \tag{5.21}$$

All currents are made up of individual electrons, but not all currents exhibit shot noise of the form we have derived. We made the fundamental assumption that the individual events (photon arrivals, or photoelectron creations) are independent of one another. In the semi-classical view of light we are taking here, independence of photon arrivals is a consequence of the superposition principle for electromagnetic fields. For the photoelectrons, it depends on absence of saturation effects.

Shot noise is a fundamental consequence of a quantum, rather than continuum, picture of the world. Yet the only feature of quantum mechanics we used was the existence of photons. At some level, it is surprising that we can get a correct answer using a semi-classical view of the world, instead of the full apparatus of quantum mechanics. It is even more surprising when we consider that a standard quantum mechanical treatment tempts one to get the wrong answer for the radiation pressure noise. Consider the following passage from Dirac's *The Principles of Quantum Mechanics*[75]:

> Suppose we have a beam of light which is passed through some kind of interferometer, so that it gets split up into two components and the two components are subsequently made to interfere. We may ... take an incident beam consisting of only a single photon and inquire what will happen as it goes through the apparatus. This will present to us the difficulty of the conflict between the wave and corpuscular theories of light in an acute form.

Corresponding to the description that we had in the case of the polarization, we must now describe the photon as going partly into each of the two components into which the incident beam is split.

...Some time before the discovery of quantum mechanics people realized that the connexion between light waves and photons must be of a statistical character. What they did not clearly realize, however, was that the wave function gives information about the probability of *one* photon being in a particular place and not the probable number of photons in that place. The importance of the distinction can be made clear in the following way. Suppose we have a beam of light consisting of a large number of photons split up into two components of equal intensity. On the assumption that the intensity of a beam is connected with the probable number of photons in it, we should have half the total number of photons going into each component. If the two components are now made to interfere, we should require a photon in one component to be able to interfere with one in the other. Sometimes these two photons would have to annihilate one another and other times they would have to produce four photons. This would contradict the conservation of energy. The new theory, which connects the wave function with probabilities for one photon, gets over the difficulty by making each photon go partly into each of the two components. Each photon then interferes only with itself. Interference between two different photons never occurs.

We used semi-classical reasoning to predict that there should be a radiation pressure contribution to the noise in an interferometer. But if we interpret Dirac's words naively, then it would seem that there could be only correlated fluctuations of radiation pressure in the two arms. Thus there would be no interferometer noise from radiation pressure fluctuations (since the interferometer output is sensitive only to differences between the arms). With only shot noise in the problem, we could always, in principle, increase P_{in} without limit. Thus there would be no quantum limit.

We seem to have a paradox, predicting the violation of quantum mechanics by using quantum mechanics. Some physicists, for a time, defended the interpretation that there was no quantum measurement limit in an interferometer. (Weiss refers to a "lively but unpublished controversy" in the community.[76]) On the other hand, the same section of Dirac's work includes the following statement:

Let us consider now what happens when we determine the energy in one of the components. The result of such a determination must be either the whole photon or nothing at all. Thus the photon must change suddenly from being partly in one beam and partly in the other to being entirely in one of the beams. This sudden change is due to the disturbance in the translational state of the photon which the observation necessarily makes. It is impossible to predict in which of the two beams the photon will be found. Only the probability of either result can be calculated from the previous distribution of the photon over the two beams.

One could carry out the energy measurement without destroying the component beam by, for example, reflecting the beam from a movable mirror and observing the recoil. Our description of the photon allows us to infer that, *after* such an energy measurement, it would not be possible to bring about any interference effects between the two components. So long as the photon is partly in one beam and partly in the other, interference can occur when the two beams are superposed, but this possibility disappears when the photon is forced entirely into one of the beams by an observation. The other beam then no longer enters into the description of the photon, so that it counts as being entirely in the one beam in the ordinary way for any experiment that may subsequently be performed on it.

So it seems that allowing for the possibility of measuring recoil may allow our semi-classical intuition to have some truth to it after all. We don't actually measure the recoil of the masses, so we don't cause the sort of "collapse of the wave function" to which Dirac's remark refers. But allowing the masses to recoil, with a subsequent observable effect on the net interferometer phase, does make our situation more subtle than the simple case in our first selection from Dirac's book.

A clear explanation of how to understand an interferometer's quantum limit was provided by Caves in 1980.[77] He showed that a proper quantum mechanical treatment of photon shot noise in an interferometer required consideration of the possibility of noise (in the form of *vacuum fluctuations* of the electromagnetic field) entering the interferometer from the output port. When this is treated properly, then we can construct a complete and consistent quantum mechanical description of the optical read-out noise. Such a picture justifies the semi-classical derivation given above, in which power fluctuations are anti-correlated in the two arms.

Caves' remarkably fruitful idea not only clarified a longstanding confusion, but it also pointed the way toward manipulating the trade-off between shot noise and radiation pressure noise that leads to the quantum limit. The key idea is to inject into the output port some so-called *squeezed light*, in which the noise in the electro-magnetic field is not evenly and randomly distributed between the two conjugate quadratures of the oscillation. (Recall that quantum mechanical uncertainty relations only govern the product of the uncertainties in two *conjugate variables*. We are familiar with position and momentum of a mass as a pair of conjugate variables. The $\sin \omega t$ and $\cos \omega t$ quadratures of an oscillating electromagnetic field have the same sort of relationship.) Application of squeezed light holds out the prospect of shifting noise from intensity noise to radiation pressure noise, thus for example allowing one to achieve the quantum limit at high frequencies with less power than would otherwise be required. Some possibilities for actually surpassing the quantum limit have also been discussed. For further information on this topic, the reader should consult the chapter by Brillet *et al.* in Blair's *The Detection of Gravitational Waves*.[78]

5.4 The Remarkable Precision of Interferometry

We cannot leave this discussion of the fundamental measurement noise without remarking on how very finely we can distinguish motion of a macroscopic mass using interferometry. Far from being limited to a precision of order a tenth or so of a fringe, we have seen that the ultimate limit to interferometer readout precision may be pushed to 10^{-9} of a fringe, or perhaps even finer. Indeed, for signals in the vicinity of 1 kHz, measurements better than $10^{-6}\lambda$ are quite routine,[6] and $10^{-8}\lambda$ sensitivity has been demonstrated in special instruments.[79] In Chapter 12, we will examine in more detail some of the practical techniques required to attain this sensitivity.

Note that this means that position noise can easily be smaller than a nuclear diameter in one second of integration time. How is this possible? Is the position of a mass even well-defined at this level of precision? In fact it is defined to a precision much finer than atomic dimensions, because the wavefront of the reflected light beam has its phase determined by the average position of all of the atoms across the beam's width. Small irregularities contribute to scattering of a small fraction of the light out of the beam, but otherwise the effects of atom-scale irregularities are negligible. (We will discuss the optics of finite-width beams in Chapter 6.)

The mass does vibrate, though; at a minimum, it must shake with an energy of $k_B T$ in its internal normal modes. (The quantum zero point energy $\frac{1}{2}\hbar\omega$ is actually much smaller, by many orders of magnitude, than that associated with Brownian motion at room temperature.) This will be one of the topics of Chapter 7. It may take some care to keep Brownian motion noise smaller than the optical readout noise of the interferometer.

Folded Interferometer Arms   **6**

Achieving optimum sensitivity to gravitational waves will require that it take light one half period of the wave to traverse an arm of an interferometer, as we saw in Chapter 2. For a wave with a frequency of about 200 Hz the required time is about 3 msec, in which case the required total (round trip) optical path length in an arm is 1000 km. The impracticality (i.e., the prohibitively high cost, among other problems) of building a device 500 km across has led to the development of schemes for folding a long optical path into a shorter length.

This need is as old as interferometry itself. Michelson and Morley confronted the same issue in constructing the instrument they used to search for the motion of the Earth with respect to the ether. Of course, the scale of their instrument was smaller, and since their signal was nearly at zero frequency they were working far from any optimum.

In modifying the original instrument of 1881, the single flat mirror at the far ends of each arm was replaced by 4 mirrors at each end of each arm. Light took a zig-zag path through the arm, traversing a total path of 22 meters in a single arm, on a slab 1.5 meters square. It is no surprise that one of the critical steps described in the 1887 paper is the procedure for aligning all these mirrors.

It is possible that the best way to fold a path many times is in fact to use many mirrors separately aligned. But several other schemes have been devised, with the aim of making alignment simpler, or with making the arm more compact in its transverse dimensions. It is important to remember one other important difference between the ether drift interferometer and a gravitational wave interferometer — in the latter, it is crucial that the mirrors be mounted on free masses. So the alignment is not something that is found once and bolted down to a rigid frame, but must be continually maintained by *control systems* that monitor any misalignments and apply small correction forces. Any reduction in the number of independent degrees of freedom is a great virtue.

In this chapter, we'll discuss two optical arrangements, both of which have been used in laboratory prototypes and proposed for multi-kilometer interferometers. The first, the *Herriott delay line*, is a direct descendant of the Michelson and Morely scheme; it uses two large curved mirrors in place of the many small ones. The second, the *Fabry-Perot cavity*, uses only two small mirrors; the light beams are not spatially separated, but are trapped for many round trips between not-quite-perfectly reflecting mirrors.

6.1 Herriott Delay Line

To introduce the delay line constructed from two curved mirrors, it is pedagogically worthwhile to first consider something that looks rather different: a semi-infinite optical "waveguide" consisting of a chain of identical lenses of focal length f separated along the x axis from one another by a distance L. (See Figure 6.1). Imagine that a narrow beam (or ray) of light is injected at a point a distance L behind the first lens: in this plane, specify the input ray by its y and z coordinates and by its slopes dy/dx and dz/dx. It propagates along a straight path to the first lens, where it is redirected by refraction. The light travels to the next lens, and the process is repeated down the chain of lenses.

The question to ask is, "Can one arrange for any simple patterns to result from a system with so many degrees of freedom?" The answer turns out to be "Yes."[80] As long as the spacing between the lenses is less than four times the focal length, it is not hard to arrange a choice of injection coordinates and angles so that the light will always encounter the lenses at a distance r_0 from the center. On each successive encounter, the ray's coordinates rotate an angle θ about the $\hat{x}$ axis from its intersection coordinates with the previous lens. For example, one could choose to inject light at a point $(y, z) = (y_0, 0)$, with angles given by

$$\frac{dy}{dx} = -\frac{y_0}{2f} \tag{6.1}$$

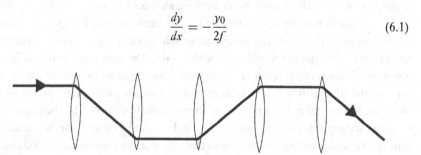

Figure 6.1: A schematic diagram of an optical waveguide constructed from lenses.

and

$$\frac{dz}{dx} = \frac{y_0}{2f}\sqrt{\frac{4f}{L} - 1}, \qquad (6.2)$$

where L is the separation of the lenses and f is the focal length of each lens. The angle θ is then given by

$$\cos\theta = 1 - \frac{L}{2f}. \qquad (6.3)$$

It is, furthermore, possible to choose the spacing L between successive lenses (or their focal lengths f) so that the light beam returns to the transverse coordinates $(y_0, 0)$ at lens number $2\mathcal{N}$. The condition for this to be true is that

$$2\mathcal{N}\theta = 2\pi\mathcal{M}, \qquad (6.4)$$

where $\mathcal{M}$ is some integer. Clearly, there are many such values of L (or f), each giving a different integer $\mathcal{N}$.

This is pretty, but its utility is only revealed if we imagine replacing the semi-infinite chain of lenses with a single pair of concave mirrors of separation L and focal length f, that is, with radius of curvature $R = 2f$. (See Figure 6.2.) A small hole in one of the mirrors allows light to be injected into the space between the two mirrors. The path down the lens waveguide has thus been transformed into a folded path between the two mirrors. If the separation L (or radius R) is chosen so that the light returns to the injection coordinates after $\mathcal{N}$ round trips, then the light will emerge from the injection hole having traveled a path of approximately $2\mathcal{N}L$. It is not hard to show that it emerges at the angle at which it would have been reflected from the input mirror, had the input hole not been drilled out of the mirror.

From this description, it might appear that such a *Herriott delay line*[81] is a tricky thing to align. But in fact, alignment is relatively simple. In effect, most of the alignment burden has been passed on to the fabricator of the two curved mirrors,

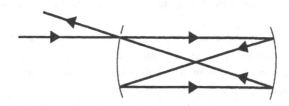

Figure 6.2: A schematic diagram of a Herriott delay line.

each of which takes the place of a whole set of small mirrors. Herriott and his colleagues invented this configuration as part of research at Bell Labs on optical signal processing.

What makes this useful is that, as long as we are close to the *reentrant condition*, as this alignment is called, the behavior of the total optical path length is a very simple function of the coordinates of the mirrors. The incremental change of optical path length with axial displacement Δx of one of the mirrors, is just $2\mathcal{N}\Delta x$. And, happily, the optical path length only depends in second order on transverse displacements of the mirrors or on rotations. (That is, the leading term in the approximation goes as $(\Delta y)^2/\ell_y$ or $\ell_\theta(\Delta\theta)^2$ where ℓ_y and ℓ_θ are lengths of order L that depend on the particular values of R and L.)[82] As long as alignment is close to right, the delay line has only one coordinate relevant to the phase of the emerging light ray, its length. Thus we have an optical system made of real components that behaves to an excellent approximation as a one-dimensional optical system. This simplifies life enormously.

6.2 Beam Diameter and Mirror Diameter

In the preceding discussion, we've treated light as a ray phenomenon, ignoring the transverse dimensions of the beam itself. But recall that Michelson and Morley's lamp gave them a beam of substantial physical extent, across which they observed interference fringes. Obviously, the size of the beam determined the appropriate size for the mirrors of their interferometer.

One of the major advantages of a laser as a light source is its ability to provide a large amount of optical power in a very narrow beam. For many purposes, over laboratory dimensions, this makes a very good physical approximation to an ideal ray. But even a laser beam cannot have an infinitesimal width, because the propagation of light over large distances involves the phenomenon of *diffraction*. When we consider propagation of light over paths many kilometers in length, we will see that it is not possible to ignore the finite width of our beams. Diffraction effects are crucial in determining the width of the beam, and thus the minimum size of mirrors for an interferometer.

We can treat diffraction in an especially simple way if we consider *Gaussian beams*, that is beams with Gaussian amplitude (e.g. electric field) profiles. Diffraction can be thought of as a process by which the amplitude distribution of light in one plane is transformed into a far-field distribution that is the original distribution's spatial Fourier transform. Recall that the Fourier transform of a Gaussian is itself another Gaussian. So, a Gaussian beam will keep its Gaussian form as it propagates through free space.

The solutions of the scalar wave equation (for, say, one of the electric field components in the wave) are of the form[83]

$$u_{00}(x,y,z) = \sqrt{\frac{2}{\pi}}\frac{1}{x}e^{i\phi}e^{-(y^2+z^2)/w^2}e^{-ik(y^2+z^2)/2R}.$$ (6.5)

The Gaussian width w is given by

$$w^2(x) = w_0^2\left[1+\left(\frac{\lambda x}{\pi w_0^2}\right)^2\right],$$ (6.6)

while the radius of curvature is given by

$$\frac{1}{R(x)} = \frac{x}{x^2+(\pi w_0^2/\lambda)^2}.$$ (6.7)

The phase shift ϕ is specified by

$$\tan\phi = \frac{x}{\pi w_0^2/\lambda}.$$ (6.8)

This wave has, as argued above, a Gaussian profile everywhere, characterized by a width $w(x)$. Near the region of minimum width (the *waist* of the beam at $x = 0$) the beam has characteristic width $w_0 \equiv \sqrt{2b/k}$, but far from this region it widens as if filling a cone of half-angle

$$\Theta = \frac{\lambda}{\pi w_0}.$$ (6.9)

This is in accord with one's expectations from diffraction problems of the more familiar variety, in which the width of an aperture is inversely related to the diffraction angle. Here, the waist of the beam functions as the effective aperture.

The foregoing encapsulates the basic physics of diffraction-limited propagation of light beams in free space. But there is a clever way to capture free waves with curved mirrors, in order to guarantee that the beam doesn't continue to spread out as it makes its many passes through a delay line. Recall that the wavefronts of the beam are characterized by a radius of curvature $R(x)$, given by Eq. 6.7. If the beam encountered a mirror with the same radius of curvature whose surface was aligned with the expanding wave front, then the direction of propagation would be simply reversed. The outgoing part of the solution becomes an ingoing part, and the wave converges to a waist again before spreading out. If the beam encountered another similarly aligned mirror on the other side of the waist, the process would repeat, returning the beam in its original direction. Thus the beam could execute many traversals of the space between the two mirrors, without very much spreading from diffraction. In effect, the curved mirror repeatedly focuses the beam to a diffraction-limited focus, the beam waist.

The situation in a Herriott delay line is similar to this, but with one important difference. All of the beams in a delay line are propagating "off axis". This means that, even if the curvature of the mirror is matched to that of the wavefront, the mirror's surface will not coincide with a single surface of constant phase. This is of course because of the tilt of that section of the mirror away from the normal to the beam's axis. It is precisely this tilt that is responsible for guiding the beam in the multi-pass pattern that we derived previously from just ray-tracing arguments. The two pictures are of course compatible: the mirror's tilt guides the beam in its repeating pattern, and its curvature focuses the beam back to a waist in the middle of the delay line.

From these considerations of diffraction-limited beam propagation, we can find the minimum allowed size for the beam cross-section at the mirrors of a delay line. Clearly, there must be a minimum size for any given delay line length L. If we choose too small a waist, the diffraction angle Θ will be large, and the beam will expand to a large width at the mirror. Too large a waist, and the beam is too wide at the mirror even though the diffraction angle Θ is small. We find the optimum by setting the derivative of Eq. 6.6 with respect to w_0 equal to zero at $x = L/2$. At the optimum, which occurs when diffraction causes the width to grow by $\sqrt{2}$ from waist to mirror, we have

$$w_{min}(L/2) = \sqrt{\frac{L\lambda}{2\pi}}. \tag{6.10}$$

If $L = 4$ km, and $\lambda = 0.5$ μm, then $w_0 = 1.26$ cm and $w(d/2) = 1.8$ cm.

The existence of a minimum beam size means there must be a minimum mirror size. Intuitively, it seems clear that the beams should not overlap, but what constitutes overlapping for a beam with a Gaussian profile is a matter of some subtlety. (See the discussion of scattered light below.) Another important consideration is that the beam should pass cleanly through the entrance hole cut in the first mirror. That hole needs to be sufficiently oversized that diffraction from its edges are unimportant. Neighboring spots should also not come too close to the hole, to guard against diffraction from the edge as well as against "cross-talk" from light leaking out the hole prematurely. A somewhat conservative design rule would be to space the spots no closer than 5 times their diameter. This makes the spot circle circumference $5\sqrt{2}w_0\mathcal{N}$, or a minimum diameter (allowing some room at the edge of the mirror outside the spot circle) of about $2.4\mathcal{N}w_0$. If we were to choose $\mathcal{N} = 36$ to make the storage time of a 4 km delay line about equal to 1.0 msec, then we need mirrors about 1 meter in diameter.

This is a substantial size, especially considering the many tight specifications that will be placed on the mirrors, including exquisite figure, surface smoothness, and reflectivity. Such a large size would also put burdens on the engineering of the mirror suspensions. Finally, the mirrors set the scale for the expensive vacuum system that must enclose the interferometer.

6.3   Fabry-Perot Cavities

From the point of view of wanting the smallest possible test masses, the ideal would be to have a folding scheme in which the mirrors were no larger than required to reflect a single diffraction-limited beam. Is there a way to fold many optical paths superimposed on top of one another? There is at least one scheme, the Fabry-Perot cavity, that can attain this ideal.

Fabry-Perot resonators can be considered the optical version of electromagnetic tuned circuits of the more familiar RF or microwave type. They have, in various forms, been extensively used in optics as a frequency-selective element for optical spectroscopy, within laser cavities to select a single line for amplification, and as narrow band filters. In the following sections, we will first discuss a generic FabryPerot resonator from a point of view in which we can neglect the transverse extent of the light beam. Subsequently, we will describe the physical implementation of a long cavity in the context of illumination by Gaussian beams.

The basic operating principle of a Fabry-Perot resonator can be understood by treating it purely as a one-dimensional system. (This is in contrast to a delay line, where the transverse position of the beam expresses the fundamental structure of its operation.) There are two familiar physical limits in which a one-dimensional description of light propagation makes sense. One is the ray optics limit that we used in the introduction of the delay line. The other is when dealing with a plane wave of infinite transverse extent; there, the x coordinate of the various surfaces of constant phase is the only interesting physical coordinate.

But, if transverse coordinates contain no relevant information, what does? The answer is the phase of the light, which we were able to ignore in our discussion of delay lines. (Of course, phase is always important, for it is the phase shift in the light that carries information about gravitational waves.)

An ideal Fabry-Perot resonator consists of two plane mirrors, numbered 1 and 2, separated by a distance L. (For now, imagine L is small with respect to the transverse dimensions of the mirror.) Describe each mirror by specifying the ratio of reflected (optical) electric field to the incident field, its *amplitude reflectivity* r_1 or r_2, and by its *amplitude transmissivity* t_1 or t_2. Conservation of energy requires

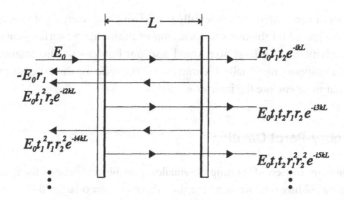

Figure 6.3: A schematic diagram of a Fabry-Perot cavity. For clarity, the spatially superposed beams have been drawn as if they were spatially separated.

that

$$r_{1,2}^2 + t_{1,2}^2 = 1, \tag{6.11}$$

as long as we can ignore losses. If not, then

$$r_{1,2}^2 + t_{1,2}^2 + a_{1,2}^2 = 1, \tag{6.12}$$

where $a_{1,2}$ is the *amplitude loss coefficient* for mirror 1 or 2. The arrangement is shown schematically in Figure 6.3. Since the mirrors to be used achieve their high reflectivities from an interface between dielectrics, with no conductors, note that there is a change of sign on reflection from a denser to a rarer medium. (This is the situation, for example, for the reflection of the incoming beam from the input mirror, since we will always want to arrange the cavity mirrors with their high-reflectivity coatings facing each other on the inboard faces of the mirrors.)

A beam of monochromatic light with electric field amplitude E_0 enters from the left, normally incident on mirror 1. Some of the light reflects from the mirror, giving a leftward moving beam with amplitude $-E_0 r_1$. The transmitted light, amplitude $E_0 t_1$, propagates to mirror 2. One beam emerges moving to the right, with complex amplitude $E_0 t_1 t_2 e^{-ikL}$, while the rest of the energy is reflected toward mirror 1. When it arrives, some is transmitted to be combined with the original reflected beam, with amplitude $E_0 t_1^2 r_2 e^{-i2kL}$. Another fraction of the light is reflected, with amplitude $E_0 t_1 r_1 r_2 e^{-i2kL}$, repeating the same trip that the first transmitted beam (of amplitude $E_0 t_1$) made to the right. There is, therefore, a series of beams transmitted from both mirrors; successive beams have complex amplitudes multiplied by additional factors of $r_1 r_2 e^{-i2kL}$.

To determine what we would actually see, we recall that all of these beams are spatially superposed, so the amplitude of the light that would reach us is the sum of the amplitudes of the individual beams. The sum is just a geometric series, so the algebra is simple; convergence is guaranteed. The field transmitted through the cavity has transmission coefficient

$$t_c = t_1 t_2 e^{-ikL} \sum_{n=0}^{\infty} (r_1 r_2 e^{-i2kL})^n = \frac{t_1 t_2 e^{-ikL}}{1 - r_1 r_2 e^{-i2kL}}. \tag{6.13}$$

The light "reflected" from the cavity, consisting of the light immediately reflected from the input mirror plus all of the beams making various numbers of round trips through the cavity, is given by the reflection coefficient

$$r_c = -r_1 + t_1^2 r_2 \sum_{n=0}^{\infty} (r_1 r_2 e^{-i2kL})^n = -r_1 + \frac{t_1^2 r_2}{1 - r_1 r_2 e^{-i2kL}}. \tag{6.14}$$

A graph of the transmitted intensity $|t_c|^2$ as a function of kL is shown in Figure 6.4. Figure 6.5 shows the phase of the reflected light.

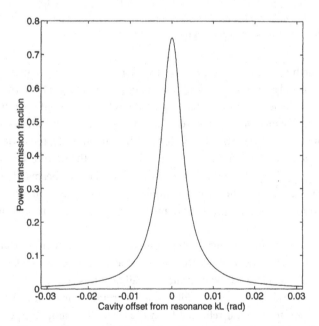

Figure 6.4: The power transmitted by a Fabry-Perot cavity in the vicinity of a resonance, as a function of the offset from resonance kL. This cavity has a finesse of 500.

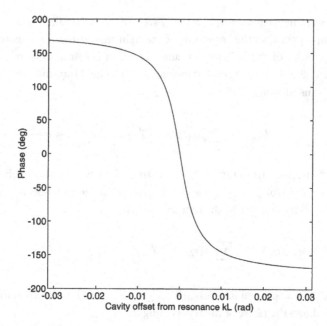

Figure 6.5:   The phase of the light reflected from a Fabry-Perot cavity in the vicinity of a resonance, as a function of the offset from resonance kL. The total phase shift from one side of the resonance to the other is 360 degrees.

It is worth trying to gain some intuition for the physical situation that we've been describing algebraically. As long as $t_2 \neq 0$, power emerges from the far end of the cavity in varying amounts, depending on the relative phase of the infinite series of beams, all of which are in some sense always present. In the situation in which the maximum power emerges ($2kL = 2n\pi$, with n any integer), the cavity is said to be in *resonance*. The amount of energy associated with the infinite series of beams within the cavity depends on the phase relations between the beams. By tuning the cavity to resonance with the light, we adjust those phase relations so that we are physically trapping power equivalent to many round trips of a single beam through the cavity.

It may seem counterintuitive that something as ghostly as a relative phase between beams can have an effect we can describe as "trapping" the light between the mirrors. We will now examine in more detail how this comes about.

For the cavities of interest for folding interferometer arms, it is worth paying special attention to the light reflected from the cavity. For simplicity, consider the special case of a perfect far mirror, that is $t_2 = a_2 = 0$, so $r_2 = 1$. In this case, the amplitude of the total reflected electric field r_c is independent of the tuning of the cavity. You can verify this algebraically by substitution into Eq. 6.14. (When we include losses or transmission of the far mirror in the model, there is a dip in intensity on resonance as shown in the Figure.)

This can be understood heuristically as follows. Far from resonance, very little light is trapped inside the cavity, so the reflection consists almost solely of the reflection of the incident beam off of the first mirror. When the cavity is tuned precisely on resonance, the electric field in the cavity is high. The amplitude of the light incident on mirror 1 from the right is (remembering we are assuming $r_2 = 1$)

$$E_{inside} = E_0 \frac{t_1}{1 - r_1 e^{-i2kL}}. \tag{6.15}$$

Only a fraction $t_1 \ll 1$ of this field escapes to be superposed with the promptly reflected field. Since resonance occurs when $e^{-i2kL} \approx 1$, or when $kL \approx n\pi$, we can give an approximate expression for the escaping electric field E_{esc}

$$E_{esc} = E_0 \frac{t_1^2}{1 - r_1(1 - i2kL)}. \tag{6.16}$$

The real part of this field is approximately equal to $E_0(1 + r_1)$. This means that

$$E_{refl} \equiv -E_0 r_1 + E_{esc} \approx E_0(-r_1 + 1 + r_1) = E_0. \tag{6.17}$$

In other words, the superposition of the escaping light with the promptly reflected light gives a beam with electric field E_0, as in the off-resonance case. But the escaping light sweeps through a phase change of a full 360° in the vicinity of the resonance. The process is illustrated in a phasor diagram in Figure 6.6.

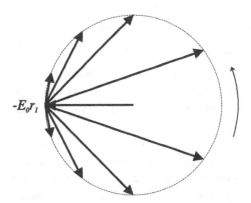

Figure 6.6: A phasor diagram of the light reflected from a Fabry-Perot cavity. The light reflected directly off the input mirror is indicated by the left-pointing phasor. Light leaving the interior gives one of the other phasors.

By examining the phase of the reflected light, we can see how similar the operation of a Fabry-Perot resonator is to the conceptually simpler delay line. In the vicinity of the resonance, the change in the phase of the escaping light as the cavity length L changes is

$$\frac{\partial \phi}{\partial L}_{FP} = \frac{4r_1}{1 - r_1} \frac{2\pi}{\lambda}. \tag{6.18}$$

Compare the equivalent expression for a delay line:

$$\frac{\partial \phi}{\partial L}_{DL} = 2\mathcal{N} \frac{2\pi}{\lambda}. \tag{6.19}$$

This change in phase with the length of the cavity is the essence of a Fabry-Perot cavity's utility as a gravitational wave interferometer folding scheme. For, just as a change in the length of the cavity can cause a phase shift near resonance, so too can a change in the appropriate metric coefficient governing the distance a light beam travels between points in the cavity. It is customary to characterize the sharpness of the resonance of a Fabry-Perot cavity by the *finesse* $\mathcal{F}$, defined as the ratio of the cavity resonance width (measured at the half power points) to the *free spectral range* $\Delta f \equiv c/2L$, the interval between resonances. For a general cavity, the finesse is given by

$$\mathcal{F} = \frac{\pi \sqrt{r_1 r_2}}{1 - r_1 r_2}, \tag{6.20}$$

which reduces to

$$\mathcal{F} = \frac{\pi \sqrt{r_1}}{1 - r_1}, \tag{6.21}$$

for the case when $r_2 = 1$ exactly. In the limit that the time scale for metric changes is long compared with $\mathcal{F}L/c$, a Fabry-Perot cavity of finesse $\mathcal{F}$ gives $2\mathcal{F}/\pi \equiv \mathcal{N}_{FP,\phi}$ times the phase shift (in response to a gravitational wave) of a one-bounce interferometer arm.

The similarities to a delay line do not end there. Light is stored in a Fabry-Perot cavity, analogously to its storage within a delay line. To see this, first imagine illuminating a well-aligned delay line for a long time, then suddenly shutting off the laser. Light continues to steadily exit the delay line for a period $2\mathcal{N}L/c$, the light travel time through the delay line pattern. It is natural to call this time the *storage time* of the delay line. Now, imagine illuminating a Fabry-Perot cavity on resonance for a long time, then suddenly shutting off the laser at $t = 0$. The "reflected" light continues to emerge, diminishing in intensity as e^{-t/τ_s} where the storage time τ_s is

$$\tau_s = \frac{L}{c}\frac{r_1}{1-r_1} \approx \frac{L}{c}\frac{\mathcal{F}}{\pi}. \tag{6.22}$$

(We have assumed here, as in Eq. 6.21, that the far mirror of the cavity is perfect; the definition has a slightly different form for a more general cavity.) Thus, although the step-function responses of the two kinds of folded arms do not have precisely the same form, each scheme can be said to store light for many round trips through the space between the mirrors. Note that for the storage time the role of the number of round trips $\mathcal{N}$ is played by $\mathcal{F}/2\pi \equiv \mathcal{N}_{FP,\tau}$, a quantity that differs by a factor of 4 from $\mathcal{N}_{FP,\phi}$, the effective number of round trips as far as phase sensitivity is concerned.

6.4 A Long Fabry-Perot Cavity

We have seen that a Fabry-Perot cavity gives a way of making light take many round trips between two mirrors while spatially superposing all of the beams. How can this be implemented in an optical system where the separation L between the mirrors is several kilometers? The considerations are rather similar to those we encountered in imagining a several kilometer long delay line. Rather than an ideal ray or ideal plane wave, we use a Gaussian beam chosen for minimum mirror size. And here, we use directly the trick of defining a cavity for beams traveling "on axis", constructing mirrors whose surfaces coincide with the phase fronts of a Gaussian beam.

The beam size in a Fabry-Perot cavity will be the same as in a delay line, hut Fabry-Perot mirrors need only be large enough to accommodate a single Gaussian beam (with the appropriate safety factor to reduce diffraction losses to tolerable levels.) Thus, a Fabry-Perot cavity makes good on its basic promise. For $L = 4$ km, the Gaussian width of the beam might be 3.6 cm, and a suitable mirror diameter around 20 cm.

6.5 Hermite-Gaussian Beams

It might occur to you to ask about illuminating a Fabry-Perot cavity with something other than a perfectly aligned beam of the proper Gaussian profile. It turns out that a pure Gaussian beam is not the only normal mode of the free-space diffraction process. Some other transverse amplitude distributions also keep their functional form as they propagate, up to scaling and curvature changes. The other mode shapes consist of Gaussian profiles multiplied by a product of Hermite polynomials in the transverse coordinates y and z.[84] A Hermite-Gaussian function is defined as

$$\psi_m(\xi) \equiv H_m(\xi)e^{-\xi^2/2}, \tag{6.23}$$

where H_m is the mth Hermite polynomial. The H_m are the same Hermite polynomials that give the stationary states, or mode shapes, for the wave function in the quantum mechanical solution of the simple harmonic oscillator. The lowest few are $H_0(\xi) = 1$, $H_1(\xi) = 2\xi$, and $H_2(\xi) = 4\xi^2 - 2$.

A general Hermite-Gaussian mode is given by

$$u_{mn}(x, y, z) = \frac{C_{mn}}{\sqrt{1 + x^2\lambda^2/\pi^2 w_0^4}}e^{i(m+n+1)\phi}$$

$$\times \psi_m\left(\frac{\sqrt{2}y}{w}\right)\psi_n\left(\frac{\sqrt{2}z}{w}\right)e^{-ik(y^2+z^2)/2R}, \tag{6.24}$$

where the C_{mn} are normalization constants. Since the 0th order Hermite polynomial is a constant, the pure Gaussian beam is the 00 mode of the set.

Each Hermite-Gaussian mode gives a simple solution to the self-diffraction (free space propagation) problem. But, in addition to having more complicated transverse spatial dependence, there is one other change to the form of the simple Gaussian beam given in Eq. 6.5. Notice that where the 00 mode has an overall phase factor $e^{i\phi}$, the mn mode has a factor $e^{i(m+n+1)\phi}$. This means that these other modes will, in general, acquire different relative phases as they propagate; thus, they will resonate for different wavelengths of light (or equivalently at different cavity lengths for fixed wavelength.) The *resonant wavelengths* are

$$\lambda_{mnp} = 2L\left[p + \frac{2(m+n+1)}{\pi}\tan^{-1}\frac{L\lambda}{2\pi w_0^2}\right]^{-1}. \tag{6.25}$$

(The index p counts wavelengths along the axis of the cavity.)

The Hermite-Gaussian modes constitute a complete orthonormal set of basis functions for a light beam in a cavity. This means that an arbitrary distribution can be expressed as a superposition of the Hermite-Gaussian modes of the cavity, with appropriate amplitude coefficients. This is useful, since the propagation problem has a simple solution for each mode. Those solutions are then superposed to find the solution of the general problem.

A nice application of these ideas is to the problem of preparing an input beam in a pure 00 Gaussian mode. One successful method is to shine the available beam onto the input mirror of a symmetric Fabry-Perot cavity (i.e. with $r_1 = r_2$), to be operated in transmission. Examine Eq. 6.25. As long as $\tan^{-1}(L\lambda/2\pi w_0^2)$ is sufficiently far from $\pi/2$, then none of the higher modes will be degenerate with the 00 mode. (This case will hold if we use the value of w_0 that gives minimum beam size, for example.) The light and cavity are tuned with respect to each other so that the 00 mode is in resonance. (More on how this tuning is done in Chapter 12). Then, the transmission factor of the 00 mode will be of order unity, while the other modes are suppressed.[85] A short Fabry-Perot cavity of this style is often used as a spatial pre-filter for the light illuminating a gravitational wave interferometer; in this application it is known colloquially as a *mode cleaner.*

6.6   Scattered Light in Interferometers

Our discussion of the operation of interferometers has assumed that the light beams follow a single well defined path through the optical system. If this assumption is violated, then the performance of the interferometer can be more complicated, i.e. noisier, than it would be otherwise. A serious problem of this sort was first noticed and diagnosed by the group at the Max Planck Institut.[86] They learned that a small fraction of the light injected into their delay line arms took a different path than intended. For example, light in the edges of the nominally Gaussian beam failed to exit the delay line through the hole in the inboard mirror; this light reentered the delay line, making another complete trip through the entire pattern before returning to exit at the hole. Imperfections in the mirrors' surfaces can also cause light to scatter away from the intended path through the delay line, reaching the beam splitter by a path whose length will in general be different.

How does this generate noise? Consider the net phase of the light returning to the beam splitter from one arm. This phase is not simply the phase accumulated by the beam that took the intended path through the interferometer. Rather, it is the "phasor sum" of the electric fields of all of the beams arriving at the beam splitter. Refer to Figure 6.7. There will in general be a phase difference between the main beam and a scattered beam. The net phase shift $\Delta\phi_{main}$ of the total light away from

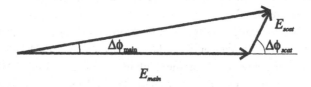

Figure 6.7:   A phasor diagram illustrating the effect of a small amount of scattered light interfering with the main beam in an interferometer.

that of the main beam is

$$\Delta\phi_{main} = \frac{E_{scat}}{E_{main}} \sin\phi_{scat}. \tag{6.26}$$

Since typically E_{scat}/E_{main} will be small, then we can expect $\Delta\phi_{main} \ll 1$. Even so, the problem is by no means negligible; remember, we are looking for gravitational wave phase shifts of the order of 10^{-9} radians or so. For this to be negligible, the power in the scattered beam would have to be less than 10^{-18}.

The problem comes from the fact that in general $\Delta\phi_{main}$ will not be constant, but will instead be noisy. For example, as the length of interferometer arm varies, so does ϕ_{scat}. At low frequencies, the "breathing" motion may amount to many wavelengths. Ordinarily we would be able to ignore this problem because it happens at low frequencies, where we will just give up on good sensitivity. But the non-linearity of the transformation, $\Delta\phi_{main} \propto \sin\phi_{scat}$, means that there will be a frequency transformation of the scattered light noise. Each time ϕ_{scat} acquires an additional 2π radians, the phase error $\Delta\phi_{main}$ oscillates through one cycle. A 1 Hz motion of 100λ in ϕ_{scat} thus shows up as a phase error $\Delta\phi_{main}$ of amplitude E_{scat}/E_{main} at a range of frequencies up to $2\pi \times 100$ Hz.

Along with diagnosing this problem in delay lines, the Max Planck group also developed a fix. The basic idea is that the scattered light adding to the main beam entered the interferometer at a different time than did the light in the main beam. This must be so if the optical paths taken by the main and scattered beams have different lengths. The scattering suppression scheme takes advantage of this fact by adding a high frequency sinusoidal phase modulation to the light entering the interferometer. (We'll describe how this can be done in Chapter 12.) The relative phase of scattered light and main beam, $\Delta\phi_{scat}$, then depends not only on the path length difference but also on the difference between the input beam's phase at the time the main beam entered and the time the scattered beam entered the interferometer. By appropriate choice of the magnitude and frequency of this modulation, it can be arranged that $\sin\phi_{scat}$ will average to zero during a modulation cycle, at least for one value of time difference between main beam and scattered beam.[86] More sophisticated phase modulation schemes have been developed that can handle more than one time difference.[87]

The delay line configuration seems especially sensitive to the scattered light problem, since its function depends on the rigorous separation of beams that snake back and forth quite close to one another both in position and in angle. Worry about this problem, both in the Herriott delay line and in a related configuration of beams known as a White cell, was one of the key factors that led Drever to propose adoption of the Fabry-Perot beam folding scheme.[88] Drever's reasoning was that with all of the folded beams already completely superposed, the worst that scattering in the arm could do was what the Fabry-Perot cavity was doing anyway. In other words, if you can't beat them, join them.

There are other places in an interferometer at which a scattered beam can interfere with the main beam. Unintended reflections from the surfaces of various optical components are another common culprit. The best strategy for dealing with them seems to be a combination of minimizing the amplitude of such beams, minimizing their overlap with the intended optical path, keeping the extra path lengths fixed, and keeping the frequency of the input light fixed so that a fixed path length difference corresponds to a fixed phase. Not all parts of the optical system are equally sensitive, of course — the crucial parts are those where the phase carries gravitational wave information (e.g. beyond the beam splitter). For large interferometers, much concern has been paid to light in the arms that scatters off the walls of the beam tube. Properly designed baffles on the walls will minimize the importance of this component of scattered light.[89]

6.7   Comparison of Fabry-Perot Cavities with Delay Lines

We closed the section on delay lines with a list of reasons to be unhappy about the large mirrors required by that optical path folding scheme. Fabry-Perot cavities solve the large mirror problem. There is one extra requirement that a Fabry-Perot mirror must satisfy; because light enters by passing through the back of a partly reflecting mirror, the mirror's body (or *substrate*) must have outstanding optical quality. The delay line at least only makes demands on the surface, since light enters through a hole.

The most important extra burden that a Fabry-Perot folding scheme has to carry is the additional complexity of having to work on resonance. Although it might seem prohibitively tricky to a beginning student of the field, it is not a problem of principle, or even of practice for a single Fabry-Perot cavity. (We will discuss how resonances may be maintained in Chapter 12). The multi-cavity systems that make up actual interferometers are decidedly complex, however. But the extra complexity seems a fair price to pay in order to attain the advantages of smaller mirrors.

6.8   Optical Readout Noise in Folded Interferometers

The point of studying optical folding schemes is to see whether one can obtain the sensitivity advantage of long optical path length without the inconvenience (or impossibility) of constructing an interferometer whose test masses are separated by 500 km or more. To finally convince ourselves that a folded path works as well as an unfolded one, we need to consider one of the fundamental performance measures of an interferometer, the magnitude of the power spectrum of the optical readout noise, as we fit an optical path of length L_{opt} into an arm of physical length L.

First, look at the shot noise component. The fluctuation in the brightness at the output depends only on the optical power and on the optical wavelength; these don't show any dependence on the length of the optical path, let alone on whether the optical path has been folded. A folded interferometer has an effective arm length difference noise that is lower by a factor of $\mathcal{N}$, since a shift Δx of a mirror causes the same effect as a shift $\mathcal{N}\Delta x$ in an unfolded interferometer. But the expression of the noise spectrum as an equivalent strain noise finally depends only on the total optical path length. (Recall from Chapter 2 that in the short storage time limit the phase accumulated by a light beam depends only on the total amount of time it spends under the influence of the gravitational wave.) We can rewrite Eq. 5.10 as

$$h_{shot}(f) = \frac{1}{\mathcal{N}L}\sqrt{\frac{\hbar c\lambda}{2\pi P_{in}}}. \tag{6.27}$$

Thus the shot noise power spectrum is the same for a folded optical path as for one with the same optical path length made using only a single round trip through long arms. This is the reason that folding is a solution to the problem of creating a path of optimum length within a device of reasonable size.

There is, however, a reason for restraint in the factor by which we seek to shrink the physical length of an interferometer by folding. Consider the amplitude of the radiation pressure noise spectrum. There, the noise comes from the force fluctuations on the test masses. A given force noise creates a noise *displacement* of each test mass. The translation from mass displacement to optical path length change does depend on the degree of folding; with $\mathcal{N}$ round trips, $\Delta L = 2\mathcal{N}\Delta x$.

There is another way that radiation pressure noise is enhanced in a folded interferometer. The light encounters each mirror $\mathcal{N}$ times, instead of once, so the force (and thus the displacement) of each mass is multiplied by $\mathcal{N}$. So in fact radiation pressure noise must be scaled up by $\mathcal{N}^2$.

When we consider practical interferometer arrangements, we'll see that Caves' original simplifying assumption of a beam splitter mass much heavier than the end masses isn't a very good model. Instead, we'll define each arm with two nearly equivalent masses. This will double the noise, since each mass in an arm feels the same force; they will be pushed apart by an amount twice the motion of each one.

So, when all the dust has settled, we have an expression

$$h_{rp}(f) = \frac{\mathcal{N}}{Lmf^2}\sqrt{\frac{2\hbar P_{in}}{\pi^3 c\lambda}}, \tag{6.28}$$

where one factor of $\mathcal{N}$ has been absorbed in the fact that L is $\mathcal{N}$ times smaller than the optimum length for an unfolded interferometer. This is much larger than in the unfolded case. The quantum limited sensitivity at frequency f will only be a factor of $\sqrt{2}$ larger, from including folding mirrors near the beam splitter that are equivalent to the end masses. But we now have a new expression for the power P_{opt} at which that level is reached at any given frequency. The quantum-limited sensitivity is achieved at $P_{opt} = \pi c\lambda mf^2/2\mathcal{N}^2$. This is a much lower power than for the unfolded case. At $f = 100$ Hz, with $m = 10$ kg and $\lambda = 0.545\,\mu$m, the optimum power is only 20 kW if $\mathcal{N} = 36$, while for the same interferometer at $f = 10$ Hz, the optimum power is 200 W. (Compare the numbers from Eq. 5.19 for the unfolded interferometer.)

6.9 Transfer Function of a Folded Interferometer

For simplicity, we have derived the shot noise limit of a folded interferometer in the low frequency limit. Our conclusions need a bit of amendment for signal frequencies f that are not small compared with c/L.

We showed in Chapter 2 how to calculate the transfer function of an interferometer in which the light makes a single round trip in the arms. It is reasonably straightforward to repeat this calculation for an interferometer with delay line arms, in which the light makes $\mathcal{N}$ round trips through an arm. For the sake of algebraic simplicity, we give the results only for the case of a gravitational wave incident with the optimum direction and polarization. The time required for light to make the second round trip is found using an expression identical to Eq. 2.18; the only difference is that the substitution we make for t is larger by an amount $2L/c$ than was used for the first round trip. Succeeding trips pick up additional increments of $2L/c$. In place of Eq. 2.23, we find[90]

$$\Delta\phi = h(t)\tau_s \frac{2\pi c}{\lambda}\mathrm{sinc}(f_{gw}\tau_s)e^{i\pi f_{gw}\tau_s}, \tag{6.29}$$

where $\tau_s = \mathcal{N}\tau_{rt0}$. A graph of the magnitude of this transfer function is shown in Figure 6.8.

The calculation of the transfer function for an interferometer with Fabry-Perot cavities as the path folding element is similar in principle. The algebra is more

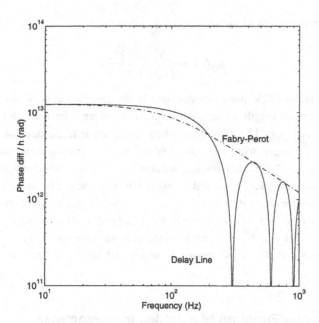

Figure 6.8: Transfer functions of interferometric gravitational wave detectors made with delay lines and with Fabry-Perot cavities for arms. The delay line's light storage time is 3.3 msec. The Fabry-Perot storage time gives matching DC response.

involved, however, so we will simply sketch the result.[91] As we saw in Section 6.3, the light emerging from a Fabry-Perot cavity is a superposition of beams having taken one, two, up to many round trips in the arm. Each of these beams has acquired the appropriate gravitational phase shift for light in a delay line of that many round trips. The net phase is obtained from the "phasor sum" of the beams. The magnitude of the transfer function can be approximated, as long as f is not too close to c/L, by the expression

$$\Delta\phi = h\tau_s \frac{8\pi c}{\lambda} \frac{1}{\sqrt{1 + (4\pi f_{gw}\tau_s)^2}}. \tag{6.30}$$

Note that this expression, like Eq. 6.29 above, is true for a wave of optimum polarization, incident along the $\hat{z}$ axis.

Examination of Eq. 6.30 reveals an important property of the Fabry-Perot transfer function. In the limit $2\pi f_{gw}\tau_{stor} \gg 1$, the transfer function becomes

$$\Delta\phi \approx h\frac{2c}{\lambda f_{gw}}. \tag{6.31}$$

The sensitivity is inversely proportional to the signal frequency in this regime, independent of the storage time. This is sometimes referred to as the *storage time limit*. Further increase in the finesse of the arm cavities neither helps, nor in the simple case hurts, the sensitivity. Note that this property is somewhat more benign than the long storage time limit in a non-resonant interferometer; there a longer storage time produces a more closely spaced set of frequencies with zero sensitivity in the transfer function, but with Fabry-Perot arms, there are no frequencies at which the transfer function goes to zero.

6.10   To Fold, or Not to Fold?

That is not quite the question posed by this chapter. There is really no choice; short of going into space or other expensive alternatives, it doesn't seem that test mass separations L longer than 3 or 4 kilometers are in the cards. Since we can only achieve optimum sensitivity to gravitational waves by making the light travel a path of hundreds to thousands of kilometers in an arm, folding is required. The real question is instead, "What price do we pay for folding the path?"

In frequency ranges dominated by shot noise, we pay no penalty at all in interferometers of the style we've been discussing, unless light lost at each mirror encounter becomes important. With the quality of mirror coatings now available, such effects wouldn't be important. However, there are a variety of clever schemes to enhance interferometer performance by using additional reflections. We will discuss some of these, such as the one called *power recycling*, later. When these are implemented, there is some advantage to minimizing the degree of folding, even in the shot noise limited regime.

But the dependence of radiation pressure noise on the degree of folding is a warning that there is a substantial price to be paid for folding in an interferometer of any design. And, at the low frequencies where radiation pressure noise would dominate shot noise, there is a large class of other noise sources expected to be even stronger. Chief among these are thermal noise (Brownian motion of the test masses) and seismic noise. These can be characterized by the power spectrum of *displacement noise* of the test masses. In the next two chapters, we'll show how to estimate and to minimize these noise sources.

There is a generic consequence of the existence of noise with this character. It appears as a strain noise h_{disp} of

$$h_{disp}(f) = \frac{2}{L} x(f), \qquad (6.32)$$

where the factor of 2 comes from the quadrature sum of the noise spectra of four independent but equivalent test masses to the total arm length fluctuation $\delta L = Lh$. Compare this to Eq. 6.27 above. The denominator of the right hand side here is the physical separation of the test masses, L, not the total path length L_{opt}. This means that the shorter we were to make L, the greater the importance of the local position noise would be, for a given L_{opt}. This is not as severe a dependence on folding as the radiation pressure noise shows ($\propto \mathcal{N}/L$). But it is a steep price, and it is what we will pay in the near term, since the magnitude of the local position noise sources is likely to remain dominant.

Thermal Noise 7

When we discussed the fundamental limits to the sensitivity of an interferometric measuring system, we were led to consider the inherent discreteness of light, and thus to touch on the quantum mechanical view of the world. An equally fundamental limit exists on the degree to which a test mass can remain at rest. This is the phenomenon usually called *thermal noise*, a generalization of Brownian motion. In a certain sense, this noise is also due to the discreteness of the world, but here it is the simple graininess of extended objects and their environment (due to the existence of atoms) that is responsible. Thus, as a thermodynamic phenomenon its magnitude depends on Boltzmann's constant k_B, instead of on Planck's constant h. (Its proportionality to $k_B T$ is what gives the effect the name "thermal noise".)

Thermal noise is a prototype of the kind of effect we referred to in the previous chapter as a *displacement noise*, which is characterized by a certain amplitude of motion of each test mass. This class of noise sources plays a crucial role in interferometer design, since (as we saw at the end of the previous chapter) it will set the limits to the degree to which an interferometer can be made compact by folding the optical path in its arms. The case of thermal noise is a good pedagogical example, but it isn't just that. As we will see, it will most likely determine the limiting sensitivity of gravitational wave interferometers in a frequency band centered on 100 Hz or so.

7.1 Brownian Motion

The classic form of thermal noise was discovered by the microscopist Robert Brown around 1828.[92] He observed a ceaseless jiggling motion of small grains of dust and pollen suspended in water. The motion was reminiscent of the activity of microorganisms, but was seen in grains made of any sort of material, whether organic or inorganic. (For some reason, Brown thought it important that the motion even was seen in dust made from a part of the Great Sphinx.) Brown's hypothesis was that the motion resulted from the action of a universal "vital force". A readable and illuminating account of Brownian motion and noise in general may be found in D. K. C. MacDonald's *Noise and Fluctuations: An Introduction*.[93]

The true source of Brownian motion was not understood until Einstein showed how it arose from fluctuations in the rate of impacts of individual water molecules on a grain. Recognizing that molecular impacts were also at the root of an explanation of the dissipation of a grain's kinetic energy as it moved through a fluid, Einstein showed that the mean-square displacement of a particle is

$$\overline{x_{therm}^2} = k_B T \frac{1}{3\pi a \eta} \tau, \qquad (7.1)$$

where τ is the duration of the observation, a is the radius of a spherical grain, and η is the viscosity of the fluid in which the grain is suspended. This was the first of many links between a *fluctuation* phenomenon, the random displacement of the particle, and a mechanism for *dissipation*, the viscosity of the water.

It is easy to overlook how sensational the understanding of Brownian motion was at the time of its elucidation. It embodied the clearest reason yet to believe in the actual physical existence of atoms, which might previously have been thought to be only hypothetical entities. This is, among other reasons, because the linkage between molecular impacts and Brownian motion allowed one to determine the value of Avogadro's number, and thus the mass of an individual atom. (There is no property like weight to endow an object with physical significance.) The many ramifications of this discovery are discussed in a beautiful set of articles by A. Perrin, reprinted as a book called *Atoms*, first published in 1913.[94] Einstein's papers on Brownian motion are also available collected in a single volume.[95]

7.2 Brownian Motion of a Macroscopic Mass Suspended in a Dilute Gas

It is instructive to examine an example of Brownian motion in a case where the physics is more directly relevant to conditions in a gravitational wave interferometer. Let's model an interferometer test mass as a rectangular plate of mass m and cross-sectional area A. (See Figure 7.1.) We'll hang the plate as a pendulum from a long wire, with resonant frequency f_0.

The space around the plate is filled with a dilute gas of pressure $p = nk_BT$, where n is the number density of molecules in the gas. Consider the limit when p is small enough that the mean free path ℓ for gas molecules is large compared to any

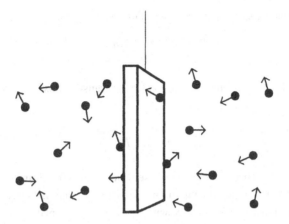

Figure 7.1: A schematic diagram of a macroscopic mass suspended in a dilute gas.

of the dimensions of the mass. Then we can neglect intermolecular collisions and concentrate only on collisions between the plate and individual molecules.

When the plate is at rest, it will be bombarded by gas molecules. On average, equal numbers arrive from each side. The rate of collisions with molecules arriving from one side is (see an introductory statistical mechanics text, such as Reif[96])

$$\mathcal{N} = \frac{1}{4} n \bar{v} A = n A \sqrt{\frac{k_B T}{2\pi\mu}}, \tag{7.2}$$

where μ is the mass of an individual molecule of the gas. Assume each of the molecules reflect elastically from the surface of the plate. The mean collision rate corresponds to a mean force from one side of just pA, or

$$F_+ = n k_B T A. \tag{7.3}$$

Pressure is the net force due to collisions from many individual molecules. When processes involve counting discrete things that arrive without correlation, we expect the fluctuations to obey Poisson statistics. In an integration time τ, we thus have a fluctuation in the total number $N\tau$ of molecules striking the plate from the left of

$$\frac{\sigma_{N\tau}}{N\tau} = \frac{1}{\sqrt{N\tau}}. \tag{7.4}$$

The fractional fluctuation in the force should be equal (perhaps up to factors of order unity) to this fractional fluctuation in the number of molecular collisions with our mass. Thus, we expect a force noise of order

$$\sigma_F^2 \sim (k_B T)^2 \frac{nA}{\bar{v}\tau}. \tag{7.5}$$

As we saw in our discussion of shot noise, we can translate this into a net force power spectrum of

$$F^2(f) \sim (k_B T)^2 \frac{nA}{\bar{v}}. \tag{7.6}$$

This is almost a useful result all by itself. But we will be rewarded if we examine the model further, to relate the fluctuating force to the frictional force the plate feels when it moves normal to its surface with velocity v_p. The plate feels a force in the direction opposite to its motion, because it receives more "backward momentum" from the molecules it sweeps up in the forward direction than it gets in "forward momentum" from the molecules trying to catch up with it from behind.

The frictional force is[97]

$$F_{fric} = -\frac{1}{4} nA\mu\bar{v}v_p \equiv -bv_p. \tag{7.7}$$

Careful comparison of this expression with Eq. 7.6 above reveals a suggestive relationship of much greater generality that this derivation would lead us to expect. With only a little algebra, we can recast Eq. 7.6 as

$$F^2(f) \sim k_B T b. \tag{7.8}$$

Here we see a basic pattern, that the force fluctuation power spectrum is proportional to the magnitude of the coefficient of dissipation.

7.3 The Fluctuation-Dissipation Theorem

We can imagine carrying out a full microphysical analysis of a number of dissipative systems, to explore how common is this relationship between fluctuation and dissipation. Fortunately, that is not necessary, because the relationship is established clearly by a theorem of great generality. It applies to any system that is linear and in thermodynamic equilibrium.

First, it is useful to make a couple of preliminary definitions. As long as a system is linear, then we can write its equation of motion in the frequency domain in terms of the amplitude of an external force $F_{ext}(f)$ necessary to cause the system to move with a sinusoidal velocity of amplitude $v(f)$. In other words, we can write the equation of motion in the form

$$F_{ext} = Zv. \tag{7.9}$$

Equivalently, we can write

$$v = YF_{ext}. \tag{7.10}$$

The function $Z(f)$ is called the *impedance*, while $Y(f) = Z^{-1}(f)$ is called the *admittance*.

These terms are useful for writing the most transparent form of the *Fluctuation-Dissipation Theorem*, as derived by H. B. Callen and his co-workers in 1951–2.[98] The theorem states that the power spectrum $F^2_{therm}(f)$ of the minimal fluctuating force on a system is given by

$$F^2_{therm}(f) = 4k_B T \Re(Z(f)), \tag{7.11}$$

where $\Re(Z)$ indicates the real (i.e. dissipative) part of the impedance. In an alternative useful form, the power spectrum of the system's fluctuating motion is given directly as

$$x^2_{therm}(f) = \frac{k_B T}{\pi^2 f^2} \Re(Y(f)). \tag{7.12}$$

Let's make this more concrete by considering as an example our gas-damped pendulum. First, write the equation of motion of the system, neglecting the fluctuating force, as

$$F_{ext} = m\ddot{x} + b\dot{x} + kx. \tag{7.13}$$

(Here, $k = m(2\pi f_0)^2$, is the effective spring constant of the pendulum, not to be confused with Boltzmann's constant k_B.) What we want to do is to treat the mass as if it supplies an "inertia force" resisting acceleration, in a manner analogous to the way a spring resists displacement from its equilibrium length, or to the way a dashpot resists relative motion of its ends. This is appropriate since we will want to be able to apply the method in cases where we treat the entire system as if it were a black box.

We want to find the external force F_{ext} necessary to establish a given velocity v. So the next step is to re-express all of the forces in the frequency domain, and to treat each term as if it were proportional to velocity. That is, $x = v/i2\pi f$, while $\ddot{x} = i2\pi f v$. The impedance $Z \equiv F/v$ is then the sum of the proportionality coefficients, in general a function of frequency. For this particular system, we find

$$Z \equiv \frac{F_{ext}}{v} = b + i2\pi f m - i\frac{k}{2\pi f}. \tag{7.14}$$

In this case we have the simple expression $\Re(Z) = b$ and so

$$F_{therm}^2(f) = 4k_B T b, \tag{7.15}$$

in line with our expectation, Eq. 7.8, from the kinetic theory analysis. Here, though, the theorem gives us the numerical constant directly.

If we want to find the power spectral density of the motion of the pendulum mass, we can use the definition of the impedance to write

$$v^2(f) = \left(\frac{1}{Z}\right)^2 F^2(f), \tag{7.16}$$

or, more familiarly,

$$x_{therm}^2(f) = \left(\frac{1}{i2\pi f Z}\right)^2 F^2(f). \tag{7.17}$$

A more direct way is to use the second form of Callen's fluctuation-dissipation theorem, Eq. 7.12. We calculate the admittance, $Y(f) \equiv Z^{-1}(f)$, as

$$Y(f) = \frac{1}{b + i2\pi f m - ik/2\pi f}, \tag{7.18}$$

or, rationalizing the denominator,

$$Y(f) = \frac{b - i2\pi f m + ik/2\pi f}{b^2 + (2\pi f m - k/2\pi f)^2}. \tag{7.19}$$

Then we can pick out the real part of the admittance directly, and write

$$x_{therm}^2(f) = \frac{k_B T b}{\pi^2 f^2 (b^2 + (2\pi f m - k/2\pi f)^2)}. \tag{7.20}$$

7.4 Remarks on the Fluctuation-Dissipation Theorem

1. One immediate advantage of this formalism is that one does not need to make a detailed microscopic model of any dissipation phenomenon in order to predict the fluctuation associated with it. (Getting the factors of 2 right in the kinetic theory calculation of gas damping takes a lot of care.) This is especially useful, since it shows how to treat on an equal footing thermal noise from different

sources of dissipation, including those from processes more complicated than gas damping. All that is needed is a macroscopic mechanical model specifying the impedance as a function of frequency. Furthermore, when several sources of dissipation act together, one simply includes their combined mechanical effects in one overall expression for the impedance.

2. Study of the papers of Callen *et al.* will reveal the essential unity of a wide variety of fluctuation phenomena. Perhaps the most well-known analog of Brownian motion is *Johnson noise* in a resistor R,

$$v^2(f) = 4k_B TR. \tag{7.21}$$

This was discovered after the explanation of Brownian motion by Einstein. The fact that we attribute this as a distinct discovery to Johnson,[99] and give credit to Nyquist[100] for its explanation, is a relic of the long history during which we finally came to understand the unity of thermodynamic fluctuation phenomena.

3. The fluctuation-dissipation theorem shows us that the way to reduce the level of thermal noise is to reduce the amount of dissipation, since $x^2_{therm}(f)$ is proportional to b divided by a function of frequency. Yet, it seems to conflict with another of our cherished beliefs from thermodynamics, the *Equipartition Theorem*: every quadratic term in a system's Hamiltonian will have a mean energy of $k_B T/2$. A one-dimensional harmonic oscillator has two such terms: the kinetic energy $mv^2/2$ and the potential energy $kx^2/2$. With respect to the latter term, equipartition requires

$$\overline{x^2} = \frac{k_B T}{k}. \tag{7.22}$$

This holds without regard to the strength of the dissipation.

How can these two statements be consistent? Actually, they are not in conflict, because they refer to different aspects of the fluctuation power spectrum. The equipartition theorem refers to the mean square fluctuation, or in other words to the integral of the displacement power spectrum. The fluctuation-dissipation theorem describes the power spectral density at each frequency, $x^2_{therm}(f) \propto \Re(Y(f))$. Therein lies the hidden complication. Consider the expression for the admittance of the velocity-damped harmonic oscillator, Eq. 7.19 above. The real part of the numerator is just the damping coefficient b. But the denominator has a richer dependence on b. Far from the resonance, the denominator's magnitude depends almost entirely on either k alone (at low frequencies) or on m alone (at high frequencies). But the shape of the admittance near the resonance is dominated by b; a low value of b means a large resonance peak as the term $(2\pi f m - k/2\pi f)^2$, vanishes.

So a small value of b means a small thermal noise driving term, and thus a small value of the thermal noise displacement spectral density at frequencies far from the resonant frequency. But in the vicinity of the resonant frequency, that is compensated by a large (resonant) response. The net effect is a displacement power spectrum $x^2(f)$ whose integral over all frequencies is independent of the amount of dissipation. This can be verified by direct integration of the power spectral density.[101]

Thus we ought to sharpen our statement on the lesson of the fluctuation-dissipation theorem: the way to reduce the thermal noise displacement spectral density away from any resonances is to reduce the amount of dissipation. This gives us a design criterion for the interferometer test masses — they need to be not only nearly free, but also nearly dissipationless.

7.5   The Quality Factor, Q

The quality factor, or Q, of an oscillator is a dimensionless measure of how small the dissipation is at the resonant frequency. It was originally used as a description of electrical resonators by electrical engineers. The definition is[102]

$$Q \equiv \frac{f_0}{\Delta f},$$

(7.23)

where f_0 is the resonant frequency, and Δf is the full width of the resonance peak in the frequency response of the system, measured at the level of half of the maximum power (that is, at $1/\sqrt{2} \approx 0.707$ of maximum amplitude.)

Consider the interpretation of this definition for the mechanical oscillator that we discussed in Chapter 4 as an example of a vibration filter. It has a transfer function

$$G(f) = \frac{k}{k - m(2\pi f)^2 + i2\pi fb'},$$

(7.24)

or, making its dimensionless character more obvious,

$$G(f) = \frac{f_0^2}{f_0^2 - f^2 + ifb/2\pi m}.$$

(7.25)

We will find it worthwhile to consider a more general expression for the damping term in an oscillator. If we replace $ifb/2\pi M$ with a generic function of frequency $ig_d(f)$, then we can write the oscillator transfer function as

$$G(f) = \frac{f_0^2}{f_0^2 - f^2 + ig_d(f)}.$$

(7.26)

It is not hard to show that the definition in Eq. 7.23 is equivalent to the statement that Q is equal to the ratio of the maximum amplitude at the resonance peak in the transfer function to the value of the transfer function well below the resonance, or

$$Q = \frac{f_0^2}{g_d(f_0)}. \tag{7.27}$$

For an oscillator with damping force proportional to velocity, that implies

$$Q = \frac{2\pi f_0 m}{b}. \tag{7.28}$$

We can also offer a time domain definition of Q. As we saw above (see Eq. 4.12), this oscillator has an impulse response of

$$g(\tau) = e^{-\tau/\tau_0} \frac{f_0^2}{f_d} \sin 2\pi/f_d \tau. \tag{7.29}$$

Then, Q is the number of radians of oscillation it takes for the amplitude of the free oscillation to fall by $1/\sqrt{e}$, or

$$Q = \pi f_0 \tau_0. \tag{7.30}$$

Substitution of the relation $\tau_0 = 2m/b$ gives the same answer for Q as does Eq. 7.28 above.

For an illustration of these concepts, see Figure 7.2.

7.6 Thermal Noise in a Gas-Damped Pendulum

Now we can use the jargon of the trade to restate our desire to minimize the dissipation of the motion of our test masses: we would say that we wanted to suspend the test mass with a high Q. What determines the Q of that suspension? The gas damping law given in Eq. 7.7 can be rewritten as the statement that a gas-damped pendulum has a quality factor of

$$Q = \frac{2\pi f_0 m}{\frac{1}{4} n A m \bar{v}}. \tag{7.31}$$

or numerically,

$$Q = 8 \times 10^9 \left(\frac{f_0}{1\,\text{Hz}} \right) \left(\frac{m}{10\,\text{kg}} \right) \left(\frac{10^{-6}\,\text{torr}}{P} \right) \left(\frac{400\,\text{cm}^2}{A} \right). \tag{7.32}$$

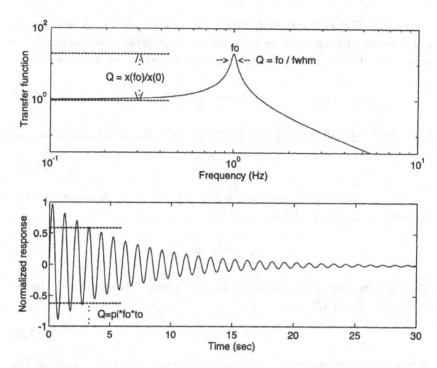

Figure 7.2: Three different manifestations of the quality factor Q of an oscillator.

Here, we have assumed that $T = 300$ K, and that the residual gas at $P = 10^{-6}$ torr is nitrogen. If the main gas component is hydrogen instead, then the prefactor in the last equation above should be 3×10^{10}, instead of 8×10^{9}.

Let's choose some sample numbers to see how important thermal noise might be. Assume we have a pendulum with a 10 kg mass at a pressure of 10^{-6} torr. Take as the resonant frequency $f_0 = 1$ Hz. The amplitude spectral density of the motion of the mass is (for frequencies well above the resonance)

$$x(f) = 2.9 \times 10^{-15} \text{ cm}/\sqrt{\text{Hz}} \left(\frac{1\,\text{Hz}}{f} \right)^2. \tag{7.33}$$

If this described a test mass in a gravitational wave interferometer with test mass separation $L = 4$ km, then we would have a dimensionless strain noise spectrum of

$$h(f) = 1.4 \times 10^{-20}/\sqrt{\text{Hz}} \left(\frac{1\,\text{Hz}}{f} \right)^2. \tag{7.34}$$

A graph of this spectral density is presented in Figure 7.3.

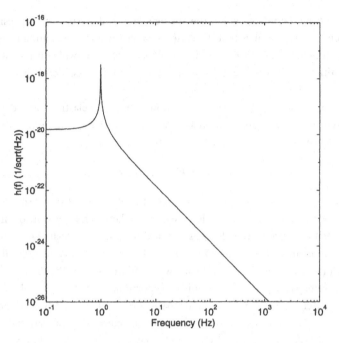

Figure 7.3: The strain noise spectrum of thermal noise due to residual gas damping. Each test mass has a mass of 10 kg; their separation is 4 km.

With good laboratory practice, there is no reason that the pressure and thus this contribution to thermal noise can't be reduced still more. Another source of velocity damping, eddy currents caused by motion of the test through the local magnetic field, can also be kept below troublesome levels, by (among other things) making the mass out of non-conducting materials.

7.7 Dissipation from Internal Friction in Materials

Unfortunately, there are other forms of dissipation that are likely to be stronger than gas damping. The limiting source of thermal noise will most likely be *internal friction* in the material of which the test mass suspension is made. A good introduction to internal friction can be found in Zener's *Elasticity and Anelasticity of Metals*.[103] To describe this phenomenon, we need to examine the departure of pendulum materials from ideal elasticity, represented by Hooke's Law

$$F_{spr} = -k\delta x. \tag{7.35}$$

Here δx is the amount by which the spring's length departs from its rest length. As written, this law implies that the elastic material responds instantaneously to an external influence. But at some level, any physical material will continue to stretch after its initial response to the application of an external stress. This behavior, known as *anelasticity*, causes dissipation of energy.

The easiest way to make the connection between anelasticity and damping is to represent it in the frequency domain. We can consider spring models that give a force of the form

$$F_{spr} = -k(1 + i\phi(f))\delta x. \tag{7.36}$$

The function $\phi(f)$, representing the degree of anelasticity of the spring, gives the phase angle in radians by which the response δx lags behind a sinusoidal driving force. If $\phi(f)$ were proportional to f, say $\phi = 2\pi\beta f$, then the imaginary term in Eq. 7.36 would be equal to $-k\beta i2\pi f \delta x = -bv$ (if $b \equiv k\beta$). Thus, if internal friction involved a force proportional to velocity, we could model it in the form of Eq. 7.36.

But, perhaps surprisingly, damping proportional to velocity is seldom a good model for internal friction in materials. Most often, $\phi(f)$ roughly independent of frequency is a much better approximation to the properties of real materials. Typical values are of order 10^{-3} to 10^{-4} for "good" materials (including most metals), and 10^{-5} to 10^{-7} for a very low loss material such as fused silica.[104] The reasons for this frequency independence of ϕ are somewhat obscure.

A good way to think about the origin of internal friction is to recognize that there are many internal degrees of freedom in a material body whose equilibrium values depend on the state of its internal stresses. Examples include the temperature, and the density of point defects, dislocations, grain boundaries, domain walls, or impurities. Typically, these internal degrees of freedom can't adjust instantaneously to a new state of stress — if there are energy barriers of height ΔE to the process by which an internal variable adjusts to its new equilibrium value, then there will be an exponential *relaxation* to the new distribution with characteristic time τ_r. Often, τ_r obeys a so-called *Arrhenius law*,

$$\tau_r = \tau_a e^{-\Delta E/k_B T}, \tag{7.37}$$

with τ_a a characteristic time between "barrier crossing attempts".

If a spring has only one important internal relaxation process, with a well-defined energy barrier ΔE, then its dynamics can be represented by an equivalent model consisting of an ideal Hooke's Law (undamped) spring k_r in parallel with a series combination of another ideal spring δk and a pure velocity-damping dashpot b. (See Figure 7.4.) The spring-dashpot combination, also known as a *Maxwell unit*,

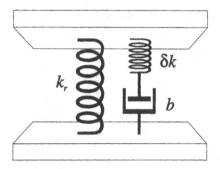

Figure 7.4: A schematic representation of a Maxwell unit, the basic building block of the description of anelasticity.

represents the contribution to the spring's stiffness of the internal degree of freedom. On time scales short compared to the relaxation time $\tau_r = b/\delta k$, the dashpot behaves as a rigid connection, and the total effective spring constant is $k_r + \delta k$. But on time scales long compared to τ_r the dashpot is effectively free, so the total spring constant is only k_r. In the frequency domain, this model spring can be described by a frequency-dependent spring constant

$$k(f) = k_r + \delta k \left(\frac{(2\pi f)^2 \tau_r^2}{1 + (2\pi f)^2 \tau_r^2} + \frac{2\pi f \tau}{1 + (2\pi f)^2 \tau_r^2} \right). \qquad (7.38)$$

A graph of real and imaginary parts is shown in Figure 7.5. The spring is lossiest where the imaginary part is greatest, a band of about one decade in width centered on $f = 1/2\pi\tau_r$ known as the *relaxation peak*. Many examples of systems exhibiting damping of this form are shown in Nowick and Berry's *Anelastic Relaxation of Crystalline Solids*.[105] If you were a materials scientist interested in, say, the mobility of defects in solids, then you'd be especially interested in the insight afforded by this simple relationship between damping and the internal degrees of freedom.

But most of the world is not so simple. The near ubiquity of materials with roughly constant ϕ was noted as early as 1927.[106] How are we to account for this, if relaxation peaks are the natural form for internal friction to take? It is not actually very hard to see the connection. First, note that a single relaxation contributes substantial damping over a decade of frequency. Next, recall that there is an exponential relation between the height of an energy barrier and the relaxation time that results. So, if there is even a modest spread in energies associated with different sites in the material, quite a large distribution of characteristic times will result, spreading a single relaxation peak into a superposition of many of them.[107] Finally, there may be many relaxation mechanisms at work in any given material. (The frontispiece

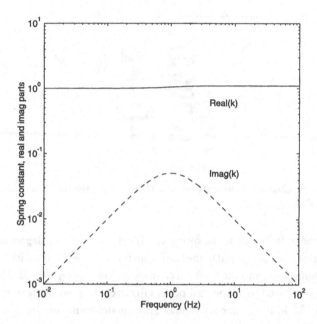

Figure 7.5: The real and imaginary parts of the the spring constant of a spring damped by a single relaxation. The relaxation is centered at 1 Hz, and has a fractional strength of 0.1.

of Zener's book[103] is a valuable graphical summary.) It has also been argued that dissipation from dislocation motion will show a frequency independent form.[108] It is almost surprising that single relaxation behavior is observable at all.

If we make an oscillator from an internally-damped spring with constant loss angle $\phi \ll 1$, its impulse response is an exponentially damped sinusoid like the one in Eq. 7.29 with $Q = \phi^{-1}$ or $\tau_0 = (\pi f_0 \phi)^{-1}$. (If $\phi(f)$ were not constant, then the value at the resonant frequency f_0 should be used in these expressions.) The admittance is similar to the expression for that shown in Eq. 7.18, except the imaginary part of the denominator is given by $ik\phi/m$. Thus, the fluctuation-dissipation theorem predicts a thermal noise spectrum for the oscillator of[109]

$$x^2(f) = \frac{4k_B Tk\phi}{2\pi f[(k - m(2\pi f)^2)^2 + k^2\phi^2]}. \tag{7.39}$$

This is quite similar to the power spectrum for the classical velocity damped oscillator (see Eq. 7.20), except that, in the constant ϕ case, the low frequency region has a $1/f$ slope instead of a constant level, and the high frequency region goes like $1/f^5$ instead of $1/f^4$.

The good news in this spectrum is the steeper slope at high frequencies. Since the resonant frequency of the test mass suspension is always going to be below the frequencies at which we are looking for gravitational waves (or else the mass wouldn't behave like a free mass), this means we are better off (by a factor of f_0/f in the power spectrum) with constant ϕ damping than with velocity damping, for a given Q of the suspension.

7.8 Special Features of the Pendulum

It is not easy to find materials whose ϕ^{-1} is even close to being as large as the Q that could be obtained if gas damping were the only form of dissipation. This would seem to say that the thermal noise of the test masses would be many orders of magnitude greater than the level we calculated in Eq. 7.33. Fortunately, though, when the *suspension* of the test mass is in the form of a pendulum, we can achieve a Q much higher than the ϕ^{-1} of the pendulum wire material itself. A pendulum is a cleverly chosen oscillator for this function. The job of keeping the mass from falling down is performed in a way quite distinct from the job of providing a horizontal restoring force. Think of the simplest possible pendulum, consisting of a single fine wire of length ℓ rigidly clamped at its bottom end to the test mass m, and at its top end to a stiff supporting structure. It is solely the elasticity of the wire that keeps the mass from falling to the floor; the wire stretches until the elastic restoring force upward balances the mass's weight. If the mass is displaced vertically, it will execute simple harmonic motion with a frequency determined by the elastic spring constant associated with the wire's stretch.

But in the horizontal direction, the story is different. When the mass is displaced from its rest position it feels a restoring force, but that force is due almost entirely to gravity, as if the wire were a rigid rod perfectly hinged at the top. That is, we get almost the right resonant frequency by using $f_0 = (1/2\pi)\sqrt{g/\ell}$. Gravity provides an effective spring constant of $k_{grav} = mg/\ell$.

There is, of course, also some restoring force from the flexing wire. It is an interesting exercise in elasticity theory to show that the fine wire under tension has a spring constant of

$$k_{el} = \frac{\sqrt{mgEI}}{2\ell^2},$$ (7.40)

where E is the modulus of elasticity of the wire material, and I is the wire cross-section's moment of inertia, as defined in elasticity theory. (See, for example, Landau and Lifshitz's *Theory of Elasticity*.[110]) As long as the wire is reasonably fine, then $k_{el} \ll k_{grav}$.

Why is this an interesting distinction to make? The reason is that only the elastic spring constant has a dissipative fraction; the gravitational spring is for all practical purposes lossless. Thus, the internal friction dissipating the energy of a pendulum's horizontal oscillation is a small fraction of the elastic spring constant, which is in turn a smaller fraction of the total effective spring constant governing the motion. This means a pendulum oscillator won't be characterized by the ϕ of the wire itself but will instead behave as an oscillator whose internal friction is given by

$$\phi_{pend} = \phi_{wire} \frac{k_{el}}{k_{el} + k_{grev}} \tag{7.41}$$

If the wire's thickness is chosen to be such that the weight of the test mass causes a stress that is an appreciable fraction of the yield stress, then k_{el}/k_{grav} can be in the range of 10^{-2} to 10^{-3} for typical materials. This means that the Q of a pendulum should be 100 to 1000 times higher than, say, a coil spring oscillator made of the same spring material as the pendulum wire. A pendulum is the least lossy way we know to make a spring.

Some rather conservative specifications for a pendulum might be: $m = 10$ kg, $f_0 = 1$ Hz, four tungsten wires with diameter of 0.12 mm and internal friction of $\phi^{-1} = 2000$. This would yield a pendulum of $Q \approx 6 \times 10^6$, and a thermal noise power spectrum of[109]

$$x(f) = 1.1 \times 10^{-13} \text{ cm}/\sqrt{\text{Hz}} \left(\frac{1 \text{ Hz}}{f}\right)^{5/2}. \tag{7.42}$$

In a gravitational wave interferometer of length $L = 4$ km, the strain noise spectrum would be

$$h(f) = 5.4 \times 10^{-19}/\sqrt{\text{Hz}} \left(\frac{1 \text{ Hz}}{f}\right)^{5/2}. \tag{7.43}$$

A graph of this spectrum is shown in Figure 7.6.

A few remarks are in order:

1. At anticipated laser power levels, thermal noise will be much more important than radiation pressure noise.
2. The graph gives an example of the general rule that shot noise dominates the interferometer noise spectrum at high frequencies (above roughly 100 Hz) while test mass position noise is important at lower frequencies. Our detailed examination of thermal noise gives one indication why this is so. The inertia of a test mass acts as a low pass filter: a spectrum of force noise $F(f)$ yields (above the suspension resonance) a spectrum of position noise $x(f) = F(f)/m(2\pi f)^2$.

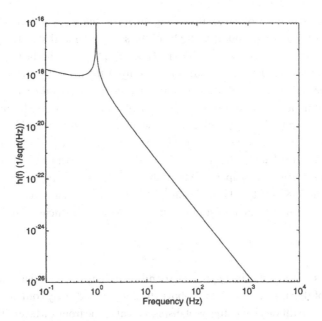

Figure 7.6: The strain noise spectral density due to thermal noise from internal friction in the pendulum wires. The parameters of this pendulum are described in the text.

In the next chapter, we will examine how to exploit this low-pass filtering effect in an aggressive way, in order to deal with a strong position noise from the environment — seismic noise.

7.9 Thermal Noise of the Pendulum's Internal Modes

The equipartition theorem reminds us that every quadratic term in a system's Hamiltonian is excited by thermal noise to a mean energy of $k_B T/2$. A real pendulum is an extended system, with many normal modes of vibration. Each of these modes can be associated with quadratic kinetic and potential energy terms. Thus, we need to consider the contribution of the thermal noise spectra of each of these modes to the total noise of the interferometer.[111]

The other relevant modes can be grouped into two classes: "violin" modes of the pendulum wires, and vibration modes of the test mass itself. The former appear as a nearly harmonic sequence with fundamental frequency

$$f_1 = \frac{1}{2}\sqrt{\frac{mg}{nm_w\ell}}, \tag{7.44}$$

where n is the number of wires supporting the mass, and m_w is an individual wire's mass. These modes will appear with Q's of the same order as that of the pendulum mode, and for the same reason.[112] Thermal noise displacement of the test mass itself is suppressed by the large mass ratio m/m_w; only at the high Q resonant peaks are these modes likely to contribute visibly, in the presence of the other noise sources.

The situation is more complicated and less benign for the modes of vibration of the test mass itself. To calculate the modal structure of a cylinder with thickness/diameter ratio of order unity involves rather complicated calculations. Fortunately, we can read about the results of an exhaustive experimental exploration of the modal structure in a paper by MacMahon.[113] The calculations weren't carried out until the 1980 work of Hutchinson.[114] For mirrors with a well-chosen ratio of thickness to diameter, the lowest modes have resonant frequencies of order

$$f_{int} = \frac{1}{2}\frac{v_s}{d}, \tag{7.45}$$

where v_s is the longitudinal speed of sound in the mirror material, and d is the diameter of the mirror. The decade above the gravest mode may contain a dozen or more modes, with varying patterns of displacement of the front surface of the mirror. (Some of these should not contribute in a well-aligned interferometer, because of the symmetry of the mode shape.)

The Q of such modes should be able to reach the inverse of the loss angle ϕ of the mirror substrate. There is, unfortunately, no "improvement factor" comparable to that applying to the pendulum mode and the violin modes. This is a crucial distinction. The only way to achieve sufficiently high Q is to find the right mirror material. Remarkably, fused silica of good purity not only has excellent optical properties, but also can exhibit $\phi < 10^{-6}$ at room temperature.[115]

The only other design parameter available for minimizing the thermal noise displacement is to strive to make the resonance frequencies as high as possible. For a compact mass appropriate to a Fabry-Perot, the lowest mode should be about 15 kHz. A mirror for an $\mathcal{N} = 36$ delay line would instead have a resonance as low as 3 Hz; however, the higher mass suppresses the motion substantially. Fortunately, in either case, the resonant frequency is high enough to be considered above the frequency range of greatest interest for astrophysical signals. We are thus mainly interested in the low frequency "tail" of the thermal noise spectrum.

The existence of these internal modes is important. It means that thermal noise doesn't just set a low-frequency "wall" to good gravitational wave sensitivity. It also will, depending on the parameters of the test masses and their suspension, set a "floor" to the sensitivity as well. The internal modes of the test masses are particularly important. We can approximate the thermal noise in the band well below all of the internal modes as

$$x(f) \approx \left(\frac{8k_B T}{2\pi f} \sum_n \frac{\phi_n(f)}{M(2\pi f_n)^2} \right)^{1/2}, \tag{7.46}$$

where the sum is over all modes n (with resonant frequencies f_n) that involve a net motion of the center of the front face of a test mass. The factor 8 in place of the usual 4 accounts for the fact that, for these modes, the displacement of the front surface is of order twice that of the mean motion in the mode. If we have a test mass with $M = 10$ kg, $\phi = 10^{-6}$, and with an aspect ratio such that the lowest two relevant modes are degenerate, then if only the lowest two modes contribute significantly, the thermal noise would be[109]

$$x(f) \approx 3.4 \times 10^{-17} \text{ cm}/\sqrt{\text{Hz}} \left(\frac{1 \text{ Hz}}{f} \right)^{1/2}, \tag{7.47}$$

assuming the mirror's ϕ is independent of frequency. This corresponds to a strain spectral density of

$$h(f) \approx 1.7 \times 10^{-22}/\sqrt{\text{Hz}} \left(\frac{1 \text{ Hz}}{f} \right)^{1/2}, \tag{7.48}$$

graphed in Figure 7.7. Note, though, that recent work by Gillespie and Raab has pointed out that the contributions of the higher modes may not be negligible; for

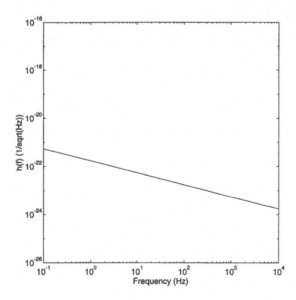

Figure 7.7: The strain noise spectral density due to thermal noise of the internal normal modes of the interferometer test masses. Parameters of the masses are described in the text.

typical test mass dimensions, the actual noise may be of order a factor of two higher than that given above.[116] This spectrum is comparable to the shot noise contribution expected in a first generation long baseline interferometer, posing an important challenge to the pace of sensitivity improvements in the frequency band near 100 Hz. Clearly this is an important topic for future research.

Seismic Noise and Vibration Isolation **8**

In this chapter, we turn our attention to another important source of displacement noise, the vibration of the terrestrial environment known as *seismic noise*. Without proper design of the instrument, seismic noise might make gravitational wave detection impossible. (Recall Michelson and Morley's difficulties caused by vibration of their instrument.) Even with careful instrument design, at some frequencies seismic noise will be the primary limit to an interferometer's sensitivity.

In spite of its crucial importance, seismic noise is not "fundamental". That is, no one knows any deep and simple physical explanation for the seismic noise spectrum. There are no fundamental constants that set the scale, as $\hbar$ does in the case of optical readout noise or k_B for thermal noise. We will take the ambient seismic noise spectrum as a given, albeit a site-specific one. Only if we could put an interferometer in space could we avoid the problem of seismic noise. (There is an ongoing effort to design space-borne interferometers for exactly this reason.[131])

"Everybody talks about the weather, but nobody does anything about it," goes the old saying. Seismic noise is the vibrational equivalent of weather. We don't know how to control its level in our environment, but we do have the ability to prevent a large portion of the seismic noise spectrum from disturbing our interferometer. That is, we can at least make a pretty good umbrella. Discussing vibration isolation techniques will form the bulk of this chapter.

8.1 Ambient Seismic Spectrum

If you were to place a calibrated seismometer on the ground (or better yet, just below ground level to shield it from the breeze) you might see noise with the following

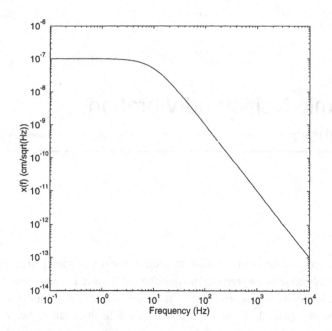

Figure 8.1: The displacement spectral density of the ground at a relatively quiet site.

spectrum:

$$x(f) = \begin{cases} 10^{-7} \text{ cm}/\sqrt{\text{Hz}}, & \text{from 1 to 10 Hz} \\ 10^{-7} \text{ cm}/\sqrt{\text{Hz}}(10 \text{ Hz}/f)^2, & \text{for } f > 10 \text{ Hz}. \end{cases} \tag{8.1}$$

(A graph of this spectrum is shown as Figure 8.1.) If you see noise substantially lower than this, you are at a very quiet site. More than likely, if you were to make the measurement where you are now, you would see noise substantially in excess of this reference spectrum. That is because in densely populated areas the shaking of the earth is dominated by vibration generated by human activities. Michelson and Morley remark that they were only able to make measurements late at night; this is a tell-tale sign that their vibration problem was generated by local human activity.

The question still arises, "What is the origin of the seismic background at sites remote from human activity?" There are a few partial answers, but no simple overall picture. In the vicinity of a period of 6 seconds, there is at most places an enhancement in the spectrum known as the *microseismic peak*. This is believed to be driven by the impacts of ocean waves on continental coastlines — it is stronger at the margins of continents than in their interiors, and shows some correlation in strength with the proximity of strong weather systems.[117]

8.2   Seismometers

A *seismometer*, the instrument used to measure seismic noise, is a very simple device. From the point of view of its dynamic response, it consists of a rigid housing that can be placed in contact with the ground, from which a *proof mass m* is suspended by means of a spring k. Typically, the resonant frequency $f_0 = (1/2\pi)\sqrt{k/m}$ is below the frequency of interest, so the proof mass stays nearly fixed in inertial space while the housing moves with the vibrating earth. Some mechanism is provided for measuring the relative displacement of housing and proof mass — their relative motion at frequencies well above the resonance is approximately equal to the motion of the earth's surface.

If we designed an instrument like this, but made the resonant frequency large compared to the frequencies of interest, then it would be called an *accelerometer* instead of a seismometer. That is because in the sub-resonant regime, the relative motion of housing and proof mass is proportional to $\ddot{x}$. Below resonance, the stiffness of the suspension, not the inertia of the mass, determines the response.

Seismometer readout can be accomplished by several techniques. A magnetic induction scheme is quite common: motion of a permanent magnet with respect to a coil generates a current proportional to their relative velocity. Various forms of capacitance bridges can be used instead — since their output is proportional to the separation of two capacitor plates, these schemes generally outperform velocity pickups at low frequencies, but are inferior at higher frequencies. Accelerometers are often constructed so that their stiff springs are made of a block of *piezoelectric* material — they supply a certain amount of electrical charge per unit acceleration.

Seismometers can use such simple readout systems because the typical seismic noise spectrum of Eq. 8.1 is so large. The cleverness involved in designing a good seismometer is not in making it sensitive enough, but in making it rugged, stable in calibration and response, insensitive to transverse motions or other influences, and in other considerations of this sort.

The large amplitude of seismic noise may be good for seismometer designers, but it is bad for gravitational wave detector designers. For one thing, it means there is no simple geophysical reason to build something with the sensitivity (and cost) of a gravitational wave detector — relatively inexpensive seismometers can see the ground moving quite well, thank you. More seriously, the large amplitude of seismic noise is a severe form of interference with the attempt to detect gravitational waves. Next we turn to a discussion of the ways in which that interference can be minimized.

8.3 Vibration Isolators

We can see how to construct a vibration isolator by just considering a seismometer from a different point of view. If you want to isolate a mass m, install it as the proof mass in a seismometer. Then, its connection with the noisy world is through a spring k that gives a resonant frequency $f_o = (1/2\pi)\sqrt{k/m}$. The equation of motion of the system is, neglecting damping for the moment,

$$m\ddot{x} = -k(x - x_g), \tag{8.2}$$

where x is the inertial coordinate of the mass, and x_g is the coordinate of the ground where the housing is attached.

In the frequency domain, we have the transfer function

$$\frac{x}{x_g} = \frac{f_0^2}{f_0^2 - f^2}. \tag{8.3}$$

This is just a simplified form of Eq. 4.22. In the limit $f \gg f_0$,

$$\frac{x}{x_g} \approx \frac{f_0^2}{f^2}. \tag{8.4}$$

This is the basic mechanism for isolation — at frequencies large compared to the resonance, the inertia of the mass prevents it from moving appreciably in response to the force kx_g before the oscillating input reverses sign. At low frequencies, on the other hand, $x/x_g \approx 1$; the spring is effectively rigid. So a vibration isolator is a low pass filter for motion, with two complex conjugate poles at f_0. A graph of the transfer function is shown in Figure 8.2.

A more complete description of how an isolator works needs to include damping as well. As we discussed in the previous chapter, there are two common idealizations of damping: velocity damping, as in a dashpot, and damping from internal friction in the spring. The transfer function in the former case is

$$\frac{x}{x_g} = \frac{f_0^2}{f_0^2 - f^2 + ifb/2\pi m}, \tag{8.5}$$

while in the latter it is given by

$$\frac{x}{x_g} = \frac{f_0^2(1 + i\phi)}{f_0^2 - f^2 + if_0^2\phi}. \tag{8.6}$$

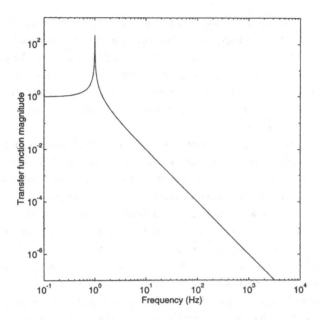

Figure 8.2: The magnitude of the transfer function of a vibration isolator whose resonant frequency is 1 Hz.

In both cases, the damping gives a bounded value for the transfer function at the resonant frequency ($x/x_g(f_0) = Q$), instead of the unphysical divergence predicted by Eq. 8.3.

There are many different kinds of mechanical realization of the required spring. They range from coils of metal wire to metal cantilevers, from blocks of rubber to a magnetic field to the wire holding the mass in a pendulum. No matter what their shape is and what material they are made of, all springs isolate a mass from vibration for the same reason: they give a soft connection to the outside world, so that there exists a resonant frequency above which the mass can't be pulled strongly enough to keep up with the motion of the outside world.

8.4 Myths About Vibration Isolation

There are many misleading hand-waving explanations of vibration isolators in the folklore of physics, a remarkable fact considering they are such simple devices. People will sometimes assert that isolation comes from vibration damping, that is from absorption of vibrational energy. Energy is absorbed if the spring constant has an imaginary part, or if a dashpot is attached to the mass. For example, a car has dashpots, called shock absorbers, in parallel with the springs that isolate the body from the vibration imparted to the wheels as they roll along a bumpy road. But the isolation has no necessary connection with the dissipation. We were able to derive the high frequency response of an isolator (its stop-band response, in filter terminology) without reference to the strength of damping present. A car's shock absorbers are there to minimize the amplification of the noise near the resonant frequency by the isolator.

Another folk explanation of vibration isolation is that it arises from a mismatch in impedance between the spring material and the mass material. This has the sound of a correct answer, and indeed it would be a relevant consideration if we were concerned with frequencies at which the acoustic wavelengths in our components were small compared with the linear dimensions of those components. Then we would be interested in how waves are reflected at the interfaces between components, a process for which the language of acoustic impedance is well suited. But this has virtually nothing to do with how a vibration isolator works, as long as it is designed with care, so that the gravest internal resonances of both the mass and the spring are at frequencies high compared to frequencies of interest. In fairness, this latter condition can sometimes be a challenging one to meet.

The impedance concept is less useful for explaining how vibration isolators work than for explaining how they fail, by the occurrence of internal standing waves, or normal modes. A dramatic illustration of the irrelevance of impedance mismatches is the carefully engineered vibration isolation system for the 50 mK resonant mass gravitational wave antenna at Stanford.[118] In it, "masses" and "springs" are alternately thick and thin sections, all made of a low-damping aluminum alloy. The isolation comes from the ability of the thick sections (the "masses") to move with respect to one another as the thin sections (the "springs") flex. There is no need for a metal-rubber interface to reflect vibrational energy. However, at the higher frequencies at which the thin sections resonate due to their own mass, vibration is conducted relatively easily through the structure.

8.5 Isolation in an Interferometer

We've previously considered the pendulum suspension of interferometer test masses as a means to approximate the condition that those test masses fall freely, and especially to do so in a way that minimizes dissipative forces. Here we meet it again from a different point of view, as an isolator from the noisy laboratory environment. (Of course, these functions are related, but they are not quite the same.)

We have a 1 Hz isolator if our pendulum is about 30 cm long. Will this provide enough isolation? Figure 8.3 shows the motion of the test mass if the top of the

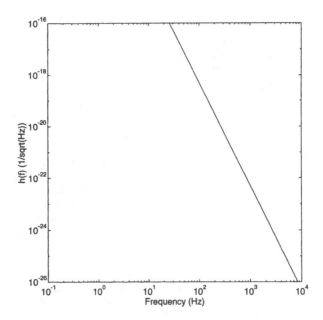

Figure 8.3:   The strain noise spectral density due to seismic noise in a 4 km interferometer, if the only vibration isolation were that provided by 1 Hz pendulum suspensions of the test masses.

pendulum wire is firmly attached to an environment with the seismic noise spectrum of Equation 8.1. Above 10 Hz, the test mass position noise spectral density is

$$x(f) = 10^{-5} \text{ cm}/\sqrt{\text{Hz}} \left(\frac{1 \text{ Hz}}{f} \right)^4, \tag{8.7}$$

equivalent to gravitational wave strain noise of

$$h(f) = 5 \times 10^{-11}/\sqrt{\text{Hz}} \left(\frac{1 \text{ Hz}}{f} \right)^4. \tag{8.8}$$

It is not until around $f = 1$ kHz that this spectrum would be low enough to search for gravitational waves of $\sim 10^{-21}$ in amplitude.

In the foregoing equations we've assumed that the pendulum behaves as an ideal one degree-of-freedom mass-spring oscillator, instead of as an extended system with many modes of vibration. In effect, we ignored the mass of the wire (the "spring"), as well as the size and elasticity of the mass. For a 1 Hz pendulum with a 10 kg mass, this is a simplification that only works well for frequencies substantially below 1 kHz or so. At higher frequencies, one needs to take account of the internal

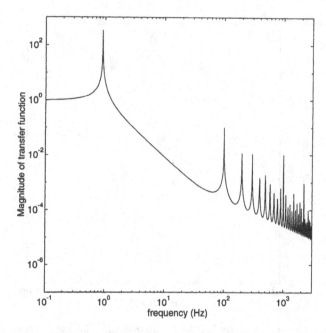

Figure 8.4: The transfer function of a pendulum, when the effect of resonances in the wires is included in the model.

resonances, especially of the wire. The transfer function will have a form given (approximately) by[119]

$$\frac{x}{x_g} = \frac{1}{\cos(2\pi f \ell/v) - [2\pi fv/g]\sin(2\pi f \ell/v)}. \tag{8.9}$$

Here, ℓ is the length of the pendulum, g is the acceleration of gravity, and v is the speed of transverse waves on the wire. A version of this transfer function is illustrated in Figure 8.4.

As we mentioned in the previous chapter, internal modes of the wires and of the test masses themselves need to be included to make a complete model of the system. For the issue of seismic noise, the important consequence of the pendulum's internal modes is that the vibration transfer function does not continue to fall like $1/f^2$ to arbitrarily high frequencies. Rather, it acquires a much shallower slope, punctuated by resonant peaks of amplitude Q above the "continuum" transmission. Even an idealized pendulum with no internal resonances provides insufficient vibration isolation by itself. That statement is all the more true when one includes the effect of the internal modes.

8.6 Stacks and Multiple Pendulums

Since we'd like to supplement the vibration isolation of the pendulum that suspends the test mass, why not suspend it from another pendulum, or indeed from a chain of many of them? It's not a bad idea, and some version of it will be part of a sensible isolation system. The basic idea is that a set of N cascaded oscillators, all with resonant frequency f_0, gives a transfer function that can be approximated, for $f \gg f_0$, as

$$\frac{x}{x_g} = \left(\frac{f_0^2}{f^2}\right)^N. \tag{8.10}$$

Proof of this is not hard, although finding a complete expression for the transfer function at all frequencies is somewhat more involved.

As good as this idea is, implementing it with a set of pendulums suffers from an important drawback, one that we've obscured by treating the question of isolation as a one-dimensional problem. A pendulum works very differently in the vertical direction than in the horizontal plane. Typically, the wire's stiffness gives a vertical resonance closer to $f_{0,vert} = 10$ Hz than the $f_{0,horiz} = 1$ Hz characteristic of the pendulum as a horizontal oscillator. This means that, above the vertical resonance frequency, a pendulum as a vertical isolator is a factor of $(f_{0,vert}/f_{0,horiz})^2$ worse than as a horizontal isolator. If we were to cascade N pendulums, we'd expect a factor of $(f_{0,vert}/v_{0,horiz})^{2N}$ between the vertical and horizontal transfer functions. With $N = 5$, this could easily be a factor of 10^{10} or more.

Why does this matter? After all, we intend to lay out the interferometer arms in a horizontal plane, precisely to take advantage of the good properties of pendulums. But it is questionable whether one could construct a chain of pendulums with sufficient symmetry that vertical noise wouldn't be converted to horizontal noise at the 10^{-3} level or so.

Even if this were possible, gravity itself guarantees that such cleverness would not be rewarded if it were installed in an interferometer with multi-kilometer arms. An arm of length L subtends at the Earth's center an angle $\alpha = L/R_\oplus$, where $R_\oplus$ is the radius of the Earth. The vertical directions at the two ends of an arm are not parallel to each other — they meet at the Earth's center. This means that a straight line can only be precisely horizontal at one point; if the arm were horizontal at one end, it would be α away from horizontal at the other end. Thus, the optic axis of the arm cannot be precisely orthogonal to the vertical; vertical noise is rejected by the orientation of the interferometer only by a factor of α. For $L = 4$ km, $\alpha \approx 6 \times 10^{-4}$. Vertical motion of the test masses would dominate the noise budget if it exceeded horizontal noise by more than $1/\alpha$.

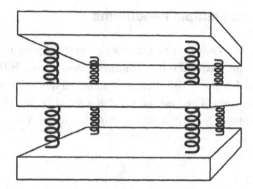

Figure 8.5:   A schematic diagram of a vibration isolation stack, constructed of metal plates separated by springs.

The vibration isolation of a pendulum is anisotropic enough so that in a chain of them the vertical noise would almost surely dominate. We are thus led to the conclusion that, while we must supplement the vibration isolation of the test mass pendulum suspension, it is almost as important to provide additional vertical isolation as horizontal isolation. Note that, coincidentally, the ratio of vertical to horizontal isolation of even a single pendulum is almost of the same order as the α appropriate to a 4 km interferometer. In other words, we need isotropic isolation, preferably a great deal of it.

This need can be met by a so-called *vibration isolation stack,* usually consisting of alternating layers of rubber springs and dense compact masses, often of lead or stainless steel. (See the schematic diagram, Figure 8.5.) Use of steel and rubber stacks for vibration isolation in gravitational wave detection was pioneered by builders of resonant mass detectors.[120] Each pair of mass-spring layers adds another pair of poles, or another factor of $1/f^2$ to the high frequency transfer function.

Stacks are quite compact, so a number of layers can be "stacked up" to provide a very steep transfer function. Rubber also has the nice property that it is well damped, so the resonances themselves don't do much amplifying of the seismic noise input. (More on the design dialectic of high Q vs. low Q in the next section.)

Stacks are certainly useful devices, but they are not simple ones. For one thing, the modal structure involves complicated couplings between the various translational and rotational degrees of freedom of the individual masses. Secondly, rubber's high degree of damping gives its elasticity a non-negligible frequency dependence. (This follows directly from the Kramers-Kronig relation.[121] See the brief discussion in Section 11.5.) Also, the internal modes of masses and springs can no more be ignored in a stack than in a pendulum. This complexity has meant that stacks were often designed by trial and error methods; now, designs usually involve finite-element modelling.[122]

We touted stacks a few paragraphs ago as a solution to the need for nearly isotropic isolation. Even for stacks with rubber springs though, it is hard to design an isotropic isolator. Rubber blocks are generally stiffer in compression (the vertical direction) than they are in shear (horizontal motion). It is not uncommon for a stack to achieve a lowest horizontal resonance frequency of 7 Hz say, but a lowest vertical resonance of only 15 Hz. So, it may still be the case that it is the vertical isolation that determines the effective position noise of the test mass, as seen by a long interferometer.

A major effort to remedy this problem was taken by the Pisa group, who built the so-called "Super Attenuator", a version of a seven layer stack in the form of seven cascaded pendulums.[123] But unlike an ordinary multiple pendulum, each stage was coupled to the next through an air spring, giving a soft compliance in the vertical direction. The parameters are such that each stage, if used by itself, would have a 1 Hz resonance frequency both horizontally and vertically. As a consequence, the isolation of the whole system is both very good and very isotropic.

The penalty that one pays is a much greater tendency to drift in height with small changes of temperature. The air in the gas springs obeys the Ideal Gas Law, so their height H tracks the temperature T like

$$\frac{\delta H}{H} \propto \frac{\delta T}{T}, \tag{8.11}$$

or a fractional height change of about $1/300$ per degree Kelvin. (The height of a solid spring depends either on the coefficient of thermal expansion or the temperature coefficient of the Young's modulus, both generally much smaller.) The Pisa system includes control of gas spring temperature to ameliorate this problem. The Pisa group is also investigating other ways to achieve good vertical compliance without gas springs. The newest version of the Super Attenuator replaces the gas springs with a set of pre-stressed cantilever springs arranged to supply a low vertical resonance frequency. The vertical compliance is softened further by the use of magnetic "anti-springs".[124]

There is one other reason to return at least partly to the conventional multiple pendulum concept. We've been implicitly assuming that the thermal noise motion of the test mass is determined entirely by the dissipation in its final pendulum suspension. But there is of course thermal noise in all parts of the isolation system. It will be especially strong if the springs are made of rubber or another very lossy material. At low frequencies, enough of this noise may be transmitted to the test mass that it may actually dominate the noise budget. (There are two equivalent ways

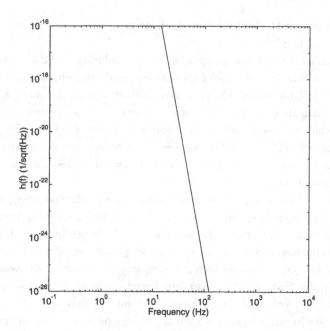

Figure 8.6:　An estimate of the strain noise spectral density in a 4 km interferometer, when the pendulum isolation is supplemented by a multi-layer stack of the sort described in the text.

to estimate this noise: either calculate the thermal noise of the last lossy stage and apply the filtering of the pendulum stage, or calculate the lossy part of the impedance at the test mass due to the lossy stack and apply the fluctuation-dissipation theorem directly. The two methods agree, as they must.)

For this reason, it may make the most sense to isolate the pendulum from the thermal noise of the stack by hanging it from one other high Q pendulum stage. That mass also provides a convenient place to which control forces for aligning the test mass can be applied. The same filtering that reduces transmitted thermal noise from the stack also filters noise from the controllers.

For reference, we show a rather conservative estimate of the influence of filtered seismic noise on an interferometer's noise spectrum $h(f)$ in Figure 8.6.

8.7　Q: High or Low?

A beginner pondering the various mechanical noise issues is sometimes tempted to ask the question, "What do you want in a mechanical isolator, high Q or low Q?"

The only correct answer to this question is "Yes." The advantages and disadvantages of each extreme need to be understood so that a successful isolation and suspension system can be designed.

We dwelled at some length in Chapter 7 on reasons why one would like a design with minimum dissipation; thermal noise is proportional to the amount of dissipation. But high Q systems come with a price — amplification of noise applied at their resonant frequencies. An oscillator with $Q \gg 1$ responds to external noise, in the steady state, with an rms amplitude of

$$x_{rms} \sim \sqrt{Qf_0} x_g(f_0). \tag{8.12}$$

This is perhaps not so bad, since the large response is concentrated at the resonant frequency. In our interferometer designs, we always try to place the resonances outside the frequency band of interest, which in this case means below the band of good gravitational wave sensitivity. If the rest of our measuring system were accurately linear, then noise at one frequency would have nothing to do with performance at another frequency, and could be ignored. We could always choose to filter it out of the output data stream, for example.

But there are limits to the range of disturbances over which a gravitational wave interferometer will behave in a linear way. Recall that the fundamental length scale in the interferometric measurement process is the wavelength of the light, about 0.5 μm for green light. This is also, in round numbers, comparable to the rms seismic noise amplitude in the octave near 1 Hz. Since a simple interferometer does not behave simply if it undergoes excursions in optical path length comparable to or larger than a wavelength, the question of the relative sizes of the linear dynamic range and the input noise amplitude needs to be faced squarely. Folding of the optical path only makes the problem worse.

This problem might seem nearly hopeless, if we didn't explain that an interferometer's linear dynamic range can be extended substantially by operating it as a feedback system in which extra control forces hold the instrument very close to a chosen operating point. We will discuss the relevant design issues in Chapter 10. For the present, it suffices to summarize the situation by remarking that feedback removes the dynamic range problem as an issue of principle, but leaves it as an extra design constraint in practice. The dynamic range concern is transferred from the interferometer itself to the feedback *actuators*, the devices that apply the control forces. These will have dynamic range limits of their own, or at least will present design trade-offs between dynamic range and some other feature, such as noise level or response rate.

The bottom line of this discussion is that it is generally best to restrict the total rms amplitude of input noise, irrespective of its frequency content. In the context of vibration isolation this gives a preference for the use of low Q resonances whenever other considerations (such as thermal noise) allow. But for the final stage of the test mass suspension, and perhaps the penultimate one as well, a high Q is the only way to keep thermal noise at a sufficiently low level. Thus, both low Q and high Q isolators have key roles to play.

In Chapter 11, we'll return to these issues; in particular, we'll examine a way to use feedback to modify the Q of an oscillator in a way that "cheats" the fluctuation-dissipation theorem.

8.8   A Gravitational "Short Circuit" Around Vibration Isolators

We emphasized above the contingent, as opposed to fundamental, nature of the seismic noise problem in gravitational wave interferometers. Similarly vibration filters, although they are governed by interesting if simple physics, are limited in their performance by what we would classify as practical, as opposed to fundamental, limitations. However, there is a rather fundamental limit to the useful amount of seismic isolation that can be applied to a test mass in a terrestrial laboratory. This comes from the direct gravitational coupling of seismic noise to test mass motion; in effect making a "short circuit" around the vibration isolation system. This effect is called *local gravitational noise* (or sometimes *gravity gradient noise*.)

When the distribution of matter in the vicinity of the interferometer is time-dependent, then the gravitational field will fluctuate as well. This will cause the test masses to move, generating a spurious gravitational response in the interferometer. Motion of macroscopic objects (such as vehicles, animals, or scientists) near the test masses can cause detectable signals in some cases.[125]

Proper instrument design and operating procedures can minimize the number of moving masses that are close enough to the test masses to matter. But there will always be a component of gravitational noise due to the seismic noise motions of the ground around the interferometer. The compressional waves in the ground cause density gradients to be established near each test mass. Transverse surface waves with vertical polarization add and subtract thin layers of material in the horizontal plane. The effects are dominated by what happens to the ground in the nearest seismic wavelength or so because of the $1/r^2$ character of the gravitational force. The coupling is, up to a factor of order unity, the same as if the test masses were attached

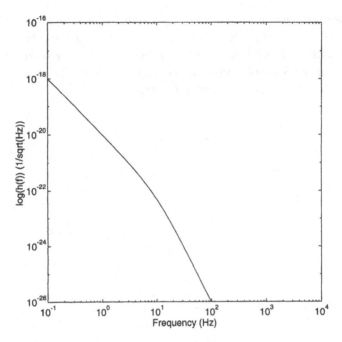

Figure 8.7: The strain noise spectral density in a 4 km interferometer due to local gravitational noise. The calculation assumes a seismic noise spectrum as given in Eq. (8.1).

to the ground with a spring that gave a resonant frequency of $f_{grav} \sim (1/2\pi)\sqrt{G\rho}$, where ρ is the density of the surface layer of the Earth near the test masses.[126]

A graph of the resulting interferometer noise spectrum, based on seismic noise of the amplitude given by Eq. 8.1, is shown in Figure 8.7. It is small in comparison with seismic noise itself, since the effective resonant frequency is so low, around 7×10^{-5} Hz. But it is important not to neglect this effect, since one can always imagine developing a very effective isolation system, such as by using a stack with many layers. Note that, if we succeeded in reducing direct seismic noise and thermal noise to levels negligible compared to the quantum limit (Eq. 5.20), local gravitational noise might still prevent us from attaining quantum limited sensitivity at frequencies lower than about 30 Hz.

8.9 Beyond Passive Isolation

For advanced interferometers, one may be interested in extending the frequency band of good gravitational wave sensitivity to a much lower cut-off frequency than is possible with passive isolation techniques. (One natural goal, adopted by the group at JILA,[127] is to reach the gravity gradient noise limit at 1 Hz.) To widen the seismically quiet band, isolation systems based on feedback control are under development. We will give a brief sketch of how they work in Chapter 11, after we have introduced the relevant control system theory.

Design Features of Large Interferometers

9

By now, we have discussed the most important physical mechanisms expected to determine the basic limits to the performance of a gravitational wave interferometer. In this chapter, we will consider two very important practical features of interferometer design. Firstly, now that we understand several important sources of position noise, we will review how one might sensibly go about choosing the overall length of an interferometer's arms. Secondly, we will examine one other feature that has a strong effect on the sensitivity of such an interferometer as well as on its cost. This is the requirement that the optical beam paths of the interferometer arms be enclosed within a high vacuum system. Finally, we will briefly describe a possible design for an interferometer in space that would make an "end run" around these concerns.

9.1 How Small Can We Make a Gravitational Wave Inteferometer?

When we examined the limits coming from the quantum nature of light, we saw that achieving sensitivity to gravitational waves of $h \sim 10^{-21}$ to 10^{-22} is possible, but only if we "push out the envelope" of interferometer technology. In particular, we saw we'd need interferometer lengths L_{opt} of order 1000 km, the maximum useful length if we are hunting for 10 millisecond long gravitational wave bursts. This is a far cry from the scale of the Michelson-Morley experiment.

But we also drew a lesson from the improvements Michelson and Morley made in going from Michelson's unsuccessful 1881 apparatus to that of 1887. Remember that an early version of his experiment had arms only 4 meters in length. When Michelson decided to extend the arms to 22 meters, he did not make an apparatus whose overall dimensions grew by a factor of six. Instead, Michelson and Morley bounced the light back and forth between nearly parallel sets of mirrors arranged on the 1.5 meter square table. To enhance the signal they sought from the ether drift, there was no requirement that the light move first only in one direction, then the opposite direction. Only the total path length mattered.

We can apply analogous reasoning to a gravitational wave interferometer. As we saw in Chapter 2, the response $\Delta\phi$ to a gravitational wave of amplitude h grows linearly with the total optical path length (at least until the light storage time becomes comparable to the gravitational wave period), just as in the ether drift experiment. Our derivation gives the same result, $\Delta\phi \propto L_{opt}$, whether the light made a single round trip between test masses separated by $L_{opt}/2$ or instead made $\mathcal{N}$ round trips in arms of length $L = L_{opt}/2\mathcal{N}$. All that matters is the total time spent moving along the $\hat{x}$ (or the $\hat{y}$) axis.

The relevant quantity to study isn't just the response $\Delta\phi$ but the signal-to-noise ratio we can achieve in a measurement. Still, as we saw in Chapters 5 and 6, if the dominant source of noise is shot noise, we come to the same conclusion. Fluctuations in the arrival rate of photons do not depend on how often the light beam reverses direction. Expressed as a strain noise $h_{shot}(f)$, the shot noise depends only on the light power and the total optical path length. So we are welcome to use the same trick as Michelson and Morley to fold the optical path to a convenient overall physical size. In Chapter 6 we discussed two folding schemes that have some advantages over the use of separate mirrors for each bounce.

But, as true as this lesson is, it is not the whole story. We needed only to consider the other face of noise from quantized light, radiation pressure noise, to see that the situation is more complicated. If we replace a simple arm of length L with one of length $L/\mathcal{N}$ containing $\mathcal{N}$ round trips, then the same radiation pressure fluctuation encounters each mirror $\mathcal{N}$ times, causing $\mathcal{N}$ times as much position noise on each test mass. Additionally, test mass displacement noise contributes $\mathcal{N}$ times as much because the light encounters each noisy mirror $\mathcal{N}$ times, instead of just once, as it makes its way through an arm. So, at low frequencies where radiation pressure noise would dominate, one has paid a high price in performance for folding the interferometer arms.

(On the face of it, radiation pressure noise is proportional to $\mathcal{N}^2$. True, but that doesn't mean the quantum limit scales the same way. That is because the quantum limit is reached at the frequency at which shot noise equals radiation pressure noise. This means a folded interferometer reaches the quantum limit at a different frequency than does an unfolded interferometer, for a given power P_{in}. When that is taken into consideration, one finds that quantum limited strain sensitivity h_{QL}

grows as a single power of $\mathcal{N}$, as if there were a fixed level of "quantum limit" position noise shaking the test masses.)

If we only had shot noise to worry about, we could fold the optical path in the arms without penalty. The existence of radiation pressure noise and the quantum limit warned us that life can't be that simple. The warning also refers to the whole class of noise sources that we've referred to as displacement noise. Many of them, including the thermal noise we studied in Chapter 7 and the seismic noise that we studied in Chapter 8, are substantially stronger than radiation pressure noise at the optical power levels that we anticipate using.

The fundamental rule is: any noise source that causes a test mass to move (with an amplitude independent of the degree of folding) will contribute $\mathcal{N}$ times as much effective strain noise in a folded interferometer as it would in an unfolded inteferometer of the same total optical path length. The existence of a multitude of such noise sources is a powerful argument against overly aggressive folding.

There are some more detailed lessons that can be drawn without considering the exact nature of the position noise spectrum. Firstly, most position noise spectra are steeply falling functions of frequency, $\propto 1/f^2$ or steeper. (We mentioned one important exception in Chapter 7, the thermal noise of test mass internal modes.) Roughly speaking, we expect that performance below 100 Hz will most likely be dominated by the struggle against test mass position noise, and thus will depend critically on $\mathcal{N}$. At higher frequencies, shot noise is likely to dominate, making 1 kHz performance less dependent on the degree to which the arms are folded. (Even there, it may be that reduction of shot noise by the recycling technique may be limited by reflection losses, and thus be sensitive to $\mathcal{N}$. See Chapter 12 for a description of recycling.)

The bottom line is that shot noise's naive independence of $\mathcal{N}$ is the exception, not the rule. The amount of path folding needs to be kept as small as possible. The stakes are illustrated in Figure 9.1, showing strain noise spectra for interferometers of the same L_{opt} but different L. How can one rationally decide on the best physical size for an interferometer? Here, inevitably, practical considerations such as cost intrude. No one has had the nerve to seriously propose construction of an unfolded interferometer 500 km in length, at least on Earth — considerations for interferometers in space are rather different. The major expensive element of a long unfolded interferometer would be the long large diameter vacuum pipe enclosing the space between the test masses (see the next section), installed in a substantial tunnel. (A straight line linking widely separated points can't follow the Earth's curved surface).

Scientific desirability seems to have met fiscal reality at a test mass separation of 3 to 4 km, corresponding to $\mathcal{N} \approx 30$ for the $L_{opt} \approx 200$ km or so appropriate for

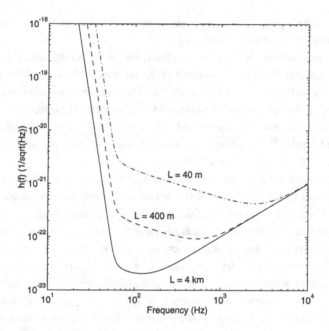

Figure 9.1: The strain noise spectral density in inteferometers with test mass separations L of 40 m, 400 m, and 4 km, respectively. The dominant noise sources are shot noise, thermal noise, and seismic noise. The parameters used are described in Chapter 16.

signals just below 1 kHz, or $\mathcal{N} \approx 300$ for observations optimized for $f < 100$ Hz. For the LIGO project, the length $L = 4$ km determines the terrain on which the battle against displacement noise will be fought.

9.2 Noise from Residual Gas

As we have emphasized, the fundamental measurable effect that a gravitational wave imparts to the light in an interferometer is the transitory shift in the relative phase of the light in the two arms. The rest of the interferometric apparatus exists to make that phase shift visible with sufficient signal-to-noise ratio and linearity. Any other physical phenomenon that can also modify the phase of a light beam is a potential source of noise.

Air at standard temperature and pressure has a profound effect on the phase of a beam of light traveling a path of $L = 1000$ km in length. The excess phase shift (compared with the same length path in vacuum) is

$$\Phi = (n_{air} - 1)L\frac{2\pi}{\lambda}, \qquad (9.1)$$

where $n_{air} \approx 1.000278$ is the index of refraction of air at $\lambda = 0.545 \ \mu$m. Of course, it is not the magnitude itself of this phase that matters, but the noise in the difference in the phase shifts in the two arms.

9.2.1 *Simple model*

How big will this noise be? We'll make a simple estimate based on Poisson ("counting") statistics for the molecules in the beam. (The derivation follows one worked out by Weiss.[129]) The model's success is based on the remarkably simple origin of the index of refraction of a dilute gas. If a single molecule of a gas has polarizability α (dimensions of length^3), then a gas of such molecules, with number density ρ has an index of refraction n of

$$n = 1 + \alpha\rho. \qquad (9.2)$$

The extension to gases containing more than one molecular species is obvious.

To estimate the excess phase shift accumulated by a Gaussian beam propagating through such a gas, we need only calculate the total number of molecules encountered by the beam. We have

$$N \approx \rho A L_{opt}, \qquad (9.3)$$

where $A \equiv \pi w_0^2 \approx L\lambda/4$ is the effective diameter of the Gaussian beam. Thus the total excess phase is about $2\pi\alpha N/\lambda A$.

The minimum noise in this number of molecules is given by the Poisson standard deviation, $\sqrt{N}$, as long as the probability for each section of the path to be occupied by a molecule is independent of the probability for all other segments. Then, since the phase noise in each arm is independent of that in the other arm, the noise in the interferometer phase is $\sqrt{2}$ times the phase shift associated with $\sqrt{N}$, or

$$\sigma_\Phi \approx \frac{2\sqrt{2}\pi\alpha}{\lambda} \sqrt{\frac{\rho L_{opt}}{A}}. \qquad (9.4)$$

Of course, we are more interested in the spectral density of this phase noise than in its rms amplitude *per se*. In any problem involving the counting of independent events (like photon shot noise in the optical read-out noise problem, or like the classic problem of shot noise in a vacuum tube diode), the power spectrum is white, up to some cut-off frequency. That cut-off represents the fact that there exists a timescale (call it τ_c) such that, on times shorter than τ_c, the number of objects we are counting is fixed. Clearly then, there can be no noise at frequencies well above $1/\tau_c$. But on time scales long compared to τ_c, each counting event is independent of any other; hence the "white" character of the power spectrum in this frequency range. Since the integral of the power spectrum over all frequencies must equal the mean square noise, the power spectral density must be of order $\sigma_\Phi^2/(1/\tau_c)$.

(N.B.: The integration time enters differently here than in the calculation of the noise from photon counting statistics, the photon shot noise. The reason is easy enough to see — there, the total number of photons we counted in an individual measurement was proportional to the integration time, since what is fixed is the rate of photon arrivals. Here, on the other hand, the number of molecules we are counting does not depend on the time scale of our measurement.)

For this problem, what is the characteristic time scale τ_c? The number of molecules in the beam changes when molecules fly into or out of the beam. The typical time that a molecule spends in the beam (call it the residence time τ_{res}) is thus the relevant time scale; it is of the order of $\tau_{res} \approx 2w_0/\bar{v}$, where w_0 is the Gaussian beam radius, and $\bar{v}$ is the mean velocity of molecules in the gas. (See Chapter 7, Eq. 7.2.) For N_2 at room temperature, in a beam with $w_o = 1.5$ cm, $\tau_{res} \approx 63$ μsec. For H_2, we have instead res $\tau_{res} \approx 17$ μsec.

Putting all this together, we predict the interferometer noise amplitude spectral density to be

$$h(f) = (\lambda/2\pi L_{opt})\Phi(f) \sim \alpha\sqrt{\frac{\rho}{L_{opt}w_0\bar{v}}}. \qquad (9.5)$$

9.2.2 *Exact result*

The estimate we've just made assumed that the interferometer uses a delay line as its beam folding scheme. This was implicit in the assumption that widely separated parts of the optical path had uncorrelated fluctuations. The situation in Fabry-Perot cavities is different. The reason is that a given wavefront traversing the interferometer arm returns to the same place before the individual molecules that were there previously have had time to "get out of the way". The net effect is to increase the phase noise because fewer independent fluctuations are averaged together.

We've also run roughshod over many subtleties, including the distribution of molecular velocities (speed and direction), and the "tapered" character of Gaussian beams. Careful consideration of these effects has been given by several (unpublished!) authors.[130] The algebra gets quite lengthy, so we will only quote the result of the calculation. The interferometer strain noise due to residual gas fluctuations is

$$h(f) = 2^{5/2}\pi^{5/4}\frac{\alpha}{\lambda^{1/4}L^{3/4}}\sqrt{\frac{\rho}{b\bar{v}}}e^{-\sqrt{2\pi\lambda L}f/\bar{v}}. \qquad (9.6)$$

In this expression, it has been assumed that the minimum beam diameter possible in an arm of length L has been used. The factor b, the number of separate beams, is $2\mathcal{N}$ for delay line arms or unity for Fabry-Perot arms. The characteristic frequency in the exponential factor is about 16 kHz if the residual gas is mainly hydrogen.

A few approximate numbers set the scale of this noise. To achieve noise of $10^{-23}/\sqrt{Hz}$ will require pressures below 10^{-6} torr if the residual gas is nitrogen, or below about 10^{-5} torr is the residual gas is hydrogen. For strain noise below $10^{-24}/\sqrt{Hz}$, 10^{-8} torr of nitrogen or 10^{-7} torr of hydrogen are the limits.

9.2.3 *Implications for Interferometer Design*

The foregoing discussion shows that, to achieve good gravitational wave sensitivity, the residual gas pressure through which the interferometer beams travel needs to be quite low. This means that the interferometer beam paths need to be enclosed by pipes over their entire length, evacuated with suitable pumps. Perhaps unlike some of the other requirements for good performance, this is a matter of standard laboratory practice, albeit on a rather large scale. An interferometer with 2 arms each 4 km long, constructed of approximately 1 meter diameter pipes, is of a different order of magnitude than a laboratory bell jar system. It is more comparable to the vacuum system of a large particle accelerator, or perhaps to a large space-simulation chamber. The requirements here are not cleverness from physicists so much as thoughtful engineering, careful construction and enough money. The vacuum enclosure and associated equipment will not only dominate the external appearance of an interferometer — they also account for a substantial share of the construction cost.

9.3 The Space-Borne Alternative

A great deal of thought has been given to the design of an interferometer for the quiet environment of space. This work has led to several well thought out ideas; we describe here one of them, the proposal for LISA, the Laser Interferometer Space Antenna.[131] This is an instrument optimized for signals in the band between 10^{-4} Hz and 1 Hz, where no conceivable terrestrial detector can operate. The proposed interferometer consists of four identical spacecraft, arranged in a V-shape of 60° opening angle, with arm length of 5×10^6 km. (Two of the spacecraft sit near each other at the vertex of the V.) The whole system is placed in a set of heliocentric orbits that have been cleverly chosen to minimize the gravitational gradients that would tend to disturb the relative positions of the spacecraft.

A Nd:YAG laser of 2 W output power is installed in each spacecraft; those in the distant craft are phase-locked to the light they receive from the vertex lasers. This active transponding is required instead of passive reflection, because optics of practical diameter (0.33 m) are so small that most light is lost to diffraction over the 5×10^6 km baseline. The shot noise is thus that appropriate to nanoWatts of power.

The space environment is mechanically quiet, but not absolutely so. Isolation from fluctuations in solar radiation pressure and solar wind is provided by constructing each spacecraft as a *drag-free satellite*.[132] This is a scheme in which a well-shielded proof mass within the spacecraft is used as an inertial reference. The rest of the spacecraft is kept centered on the proof mass by a servo system (see Chapter 11) that controls the firing of a set of thrusters. Remaining mechanical noise sources include electric and magnetic forces on the proof masses, as well as time-varying gravitational interactions between proof mass and spacecraft.

In the quietest frequency band, from 10^{-3} to 10^{-2} Hz, LISA is expected to be able to observe periodic waves of amplitude $h \sim 10^{-23}$ with integration times of 1 year. This should be sufficient to allow the observation of signals from many binary systems in the Galaxy, as well as to see a variety of massive black hole formation events from cosmological distances.

Null Instruments 10

We've been discussing the various important component parts of a gravitational wave interferometer. The emphasis has been on how those parts work as parts — how their function can influence the limiting sensitivity to gravitational waves.

But it is crucial to also think about how these parts can function together as a system. This is a subject of some depth and subtlety, toward which a great deal of cleverness has been applied in the cause of making gravitational wave interferometers work. The stark truth is that if all we did was arrange the parts we've described in the configuration of a Michelson interferometer, it would hardly work at all. If it were aligned at one moment, it most likely would not be aligned the next. If light managed to make it through the desired optical path, the output port would brighten and darken rapidly as the phase difference swept through many wavelengths. And if the arms consisted of Fabry-Perot cavities, most of the time the arms would be out of resonance, and hence "dead" — giving no appreciable phase shift in response to mirror motions, gravitational waves, or anything else.

Have we bitten off more than we can chew? Are the dual demands of ultra-low noise freely falling masses and ultra-high precision interferometric position sensing mutually incompatible? It might seem so. Interferometry is almost always carried out with a rigid apparatus. If you read catalogs for optical tables, you will see the ability to carry out interferometry cited as the ultimate proof of a table's rigidity. Michelson and Morely had to deal with vibration almost large enough to swamp the signal they were looking for. This was a problem even though the phase shift they were looking for was many orders of magnitude larger than we expect gravitational waves to be; furthermore, their experiment did not require free test masses — they bolted the mirrors as rigidly as they could to their stone slab.

Fortunately, interferometry between freely falling masses is not impossible. But it does require that the rigidifying function of an optical table be carried out in an appropriately non-rigid way. That is, the parts have to be joined into a functioning whole, while at the same time behaving more or less as we have naively described them, at least in some interesting frequency band. The set of techniques that makes this possible is called *feedback control*.

When we use feedback to control the relevant angles, positions, and lengths in an interferometer, we have at our disposal the means to do more than barely make the instrument work. By adding this extra layer of complexity, we can in fact make the response of the interferometer nearly ideally simple, and at the same time make it nearly immune to a whole class of non-fundamental noise sources. We do this by using the control system to hold the interferometer's output fixed at a well-chosen operating point, even in the presence of a signal. An instrument operated in this way is called a *null instrument*.

In this chapter, we will discuss the theory of null instruments, a subject of wide applicability throughout experimental physics. In the succeeding chapter, we will examine the rudiments of control system theory, the discipline required to realize a null instrument. Then, we will be in a position to discuss several of the most important applications of feedback to gravitational wave interferometers. The examples to which we devote Chapter 12 include control of key subsystems and of an interferometer as a whole.

10.1 Some Virtues of Nullity

The word "null" gets thrown around a lot in experimental physics. It has at least three different meanings, so it is important to distinguish between them.

10.1.1 *Null hypotheses*

In the statistical theory of signal detection, the term *null hypothesis* refers to the claim that the output of an instrument is consistent with the presence of noise alone, with no signal. This is the boring state of affairs, whose rejection would be in some way interesting. A classic example comes from radar systems. The null hypothesis might be that no attacking enemy airplane is present; it would certainly be interesting if this hypothesis were rejected. In our context we will be examining the output of our interferometer, hoping to have reason to reject the null hypothesis, "No gravitational waves passed by just now."

10.1.2 *Null experiments*

Another related but distinct meaning of the word "null" is in the concept of a *null experiment*. This refers to an experiment aimed at testing whether some quantity is precisely equal to zero. Robert Dicke championed this idea among gravitational physicists, pointing out in his book *The Theoretical Significance of Experimental Relativity* how interesting some zeroes can be in physics.[1] The difference between inertial and gravitational mass is his classic example of an interesting zero in gravitational physics. The statement that the difference should be zero is called the *Principle of Equivalence*. A zero for this difference says something very deep about physics, namely that gravity can be described by a space-time metric. It is fair to say that since Newton's time, this zero was one of the deepest puzzles in physics; Einstein's insight into the origin of this equivalence is perhaps one of the most beautiful ideas in the history of science.

Null experiments have some very important practical advantages for the experimenter. In an ordinary (non-null) measurement, the value assigned to the quantity in question depends on the calibration of the measuring instrument, its *sensitivity*. This sensitivity may vary in time unpredictably, leading to inaccuracy in the result. Furthermore, the response of the instrument may be non-linear, making accurate calibration difficult. Or, the instrument may be susceptible to some forms of noise while "deflected" to a non-zero reading that don't affect it when reading zero.

A clever experiment design is one in which the measuring instrument is in a "balanced" or "null" state when the physical quantity of interest is zero. Dicke's version of the Eötvös experiment to test the Principle of Equivalence is an archetype of such a clever style of experiment.

Let's review the basic principles of this classic experiment. If two materials, gold and aluminum say, have the same ratio of inertial to gravitational mass, then masses made from each will follow the same orbit in a gravitational field. In particular, both will feel the same gravitational attraction toward the sun in a terrestrial laboratory. If equal masses of the two materials are placed at opposite ends of a torsion balance (a horizontal beam hung from a wire), there will be no tendency for the balance to track the Sun's apparent motion through the sky. Only if the equivalence of inertial and gravitational mass were violated would the balance have a tendency to follow the Sun's attraction. Thus, the zero in the theoretical difference between the two forms of mass has been translated into a predicted zero in the response of a measuring instrument. The 1964 paper of Roll, Krotkov, and Dicke describing this experiment[133] is one of the great works of literature in experimental physics.

In case this seems so obvious that you can't even imagine another way to carry out the experiment, I offer a straightforward but fundamentally dumb Alternative Eötvös experiment. Get two cylinders, one made of gold and the other of aluminum, each of nominal mass 1 kg. Hang each one from a spring scale to determine its gravitational mass by the relation $m_{gravg} = -kx$. Then, with the mass still attached to the spring scale, swing each one around in a horizontal circular path at an angular frequency ω, to meaure the inertial mass by $m_{in}\omega^2 r = -kx$. Take the ratio m_{in}/m_{grav} for each mass, and subtract the ratios to find the difference.

What's wrong with this Alternative Eötvös experiment? Plenty. For one thing, the stiffness of the spring must be chosen to be stiff enough that the measuring device stays "on scale" during the various measurements. This dynamic range requirement will unavoidably limit the precision with which the individual mass measurements can be performed. In addition, the sequential nature of the measurement series means that any drift in the properties of the spring, say through temperature drift coupled with a non-zero temperature coefficient of the Young's modulus, will lead to a spurious indication of a violation of the Principle of Equivalence. Also, if the masses aren't exactly equal and if the angular velocities differ in the two inertial mass measurements, then any non-linearities in the spring's compliance will also give spurious violations of Equivalence.

There are probably many other things wrong with this Alternative Eötvös experiment. The point of the example is the tremendous range of possible errors in an experiment based on the subtraction of separately measured quantities. It is hard to believe one could achieve a precision of 10^{-3} in the Alternative Eötvös experiment. In the elegant Dicke version, where the experiment was cleverly designed so that the physics "performed the subtraction" before the measurement was registered, a precision of about 10^{-11} was achieved. Braginsky,[134] using similar methods, reported a null result to a precision of about 10^{-12}.

Note that, from the modern vantage point, the Michelson-Morley experiment itself would be considered a classic null experiment. Certainly, anyone repeating it today would be interested in exploring how close to zero was the difference in the speed of light traveling in different directions. The zero answer from the Michelson-Morley experiment plays the same crucial role in the Special Theory of Relativity as the Eötvös experiment's null result does in the General Theory of Relativity. This points up a bit of subjectivity in the definition of a null experiment, since Michelson (before his first attempt in 1881, in any case) would have expected a non-zero result from the experiment.

The transition of the Michelson-Morley experiment to the status of a null experiment was the occasion for one of Einstein's most famous aphorisms. In 1921, well after the triumph of not only special relativity but general relativity as well, Einstein was informed that D. C. Miller had just obtained a non-zero measurement for the ether drift. This report moved the incredulous Einstein to respond, "The Lord is subtle, but He is not malicious."[2] Compare the nearly unanimous belief of physicists in 1881 with God's physics forty years later; the contrast is striking.

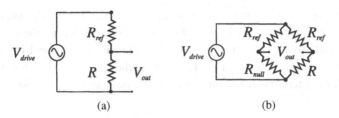

Figure 10.1: Two circuits for determining the value of an unknown resistance: (a) a voltage divider, and (b) a bridge.

10.1.3 *Null instruments*

Even when one is interested in a physical quantity that differs from zero, it is possible to achieve some of the benefits of a null experiment. This is achieved by construction of what is called a *null instrument*. The classic example of this measurement strategy is the use of a *bridge* circuit for precision electrical measurements. (See Figure 10.1b.) The component whose impedance is to be measured is placed in one leg of the circuit, in series with a well-calibrated component. The other leg consists of another pair of well-calibrated impedances, also connected in series, hooked up in parallel to the first leg. A meter measures the electrical potential difference between the mid-points of each leg. The circuit is excited with either a DC or AC voltage, depending on the application. When the impedance ratios of the pairs of components in the two legs are equal, then the meter will give a zero, or null, reading; the bridge is said to be *balanced*. But when the impedance ratios are mismatched, that fact is indicated by a non-null reading of the meter.

What happens next depends on the specific application. If the purpose of the measurement is precision impedance measurement, then usually one would adjust a calibrated variable impedance in the bridge, until the bridge is brought into the null state; the value of the unknown can be read from the calibration curve of the adjustable element. In other applications, the unknown is being used as a sensor of some quantity that modulates its impedance. Then, the bridge is usually not brought back into balance; instead, the meter reading gives an indication of the quantity to be sensed.

A bridge is the right choice for precision electrical measurements because it makes the measurement insensitive to variations in the excitation voltage, to gain drifts in the meter, or to non-linear gain in the meter. The one requirement is that the reference components in the bridge have stable impedances.

It is a useful exercise to examine how this works in a little more detail. We'll investigate how an impedance measurement might be influenced by a variation in the voltage applied to the measuring system. We'll contrast the bridge method with a simple voltage divider (see Figure 10.1a), with the latter playing the same role as did the dumb Alternative Eotvos experiment in the previous section. First, consider the simple divider. We are interested in learning the resistance of the lower resistor R, which we determine by measuring the voltage drop across it, V_{out}, and using the relation

$$R = \frac{V_{out}}{V_{drive} - V_{out}} R_{ref},$$
(10.1)

where V_{drive} is the voltage applied across the divider, and R_{ref} is the value of the upper resistor. If we assume V_{drive} is fixed when it actually varies, then we will derive an incorrect value for R from our measurement of V_{out}. In the case where $R \approx R_{ref}$, the errors in numerator and denominator are of equal magnitude but opposite signs. Thus our sensitivity to varying drive voltage is given by

$$\frac{\Delta R}{R_{ref}} \approx 2 \frac{\Delta V_{drive}}{V_{drive}}.$$
(10.2)

Now compare how a bridge behaves in response to a similar perturbation. If it were possible to null the bridge precisely, then we would have precisely zero sensitivity to the drive voltage V_{drive}. But even when a precise null is not possible, an approximate balance still gives substantial reduction in feedthrough of this kind of noise. In the general state, we would determine the value of R by measuring V_{out} and using the relation

$$R = \frac{V_{out} R_{ref} + V_{drive}(R_{ref} R_{null}/(R_{ref} + R_{null}))}{V_{drive}(R_{ref}/(R_{ref} + R_{null})) - V_{out}}.$$
(10.3)

This rather ugly expression simplifies greatly when we can assume that the bridge is close enough to balance that $V_{out} \ll V_{drive}$. For definiteness, let us also assume that $R \approx R_{ref} \approx R_{null}$. Then, a little rearrangement followed by use of the binomial expansion yields the transparent relation

$$R \approx R_{null}\left(1 + 2\frac{V_{out}}{V_{drive}}\right) + R_{ref}\left(2\frac{V_{out}}{V_{drive}}\right).$$
(10.4)

Again we would assume, erroneously, that V_{drive} was fixed, and calculate an incorrect value for R from the perturbed value of V_{out}. But now our expression for the error reads

$$\frac{\Delta R}{R_{ref}} \approx 4 \frac{V_{out}}{V_{drive}} \frac{\Delta V_{drive}}{V_{drive}} \ll \frac{\Delta V_{drive}}{V_{drive}}, \tag{10.5}$$

where the last inequality is true as long as the bridge is nearly balanced, $V_{out} \ll V_{drive}$. Thus, this simple null instrument delivers on its promise, to make the measurement nearly insensitive to an otherwise important source of noise.

10.1.4 *Null features of a gravitational wave interferometer*

A gravitational wave detector is not a null experiment in the classic sense. We do indeed hope and expect to detect gravitational waves, and so expect at some level a non-zero output from the instrument. But gravitational wave interferometers have been arranged to exhibit many features of a null instrument. Indeed, a Michelson interferometer can be considered the optical equivalent of a bridge circuit, in the degree to which it has been arranged to be insensitive to a variety of perturbing influences.

The use of two arms of equal length is a design choice made particularly so that an inteferometer can behave as a null instrument. We presented this earlier as a natural symmetrical choice to make, since a normally incident gravitational wave has equal and opposite effects on the light travel time in the two arms. And since the interference at the beam splitter measures the difference in light travel times, we get, in effect, twice the signal as if we had used just one arm. But if this were the only reason to use two equal arms, it might not justify the extra trouble. We could imagine building an interferometer with only one long arm. The second arm could be made merely vestigial, a short optical path installed simply to provide a second beam to interfere with the first. On paper this might work, and provide sensitivity down by a factor of two in exchange for only half the number of long arms.

Ordinarily, this might be an offer you could refuse, since it would cost a factor of two in sensitivity, and would probably not even cut the cost of a large interferometer in half. But what if a one-armed design would cut the cost by more than a factor of two, or otherwise offer a cost-benefit advantage, say by making available a superb site in which an interferometer of two equal arms just wouldn't fit? There are many multi-kilometer straight tunnels, including abandoned mines, but L-shaped tunnels of the requisite size are a rarity.

Unfortunately, an unequal-arm interferometer turns out to be a terrible idea. One of the most important reasons is that the interference that occurs between the two beams can be interpreted as a light travel time difference measurement (up to an integer number of optical periods) only if the optical period is strictly constant. The actual measurement involves the time difference between arrivals of optical wavefronts from the two arms. If the arms are unequal and if the laser frequency $\nu = c/\lambda$ changes substantially during the storage time of the long arm, then the wave fronts will arrive "out of sync" even if the mirrors have stayed fixed. The conversion between noise $\nu(f)$ in the laser frequency and spurious gravitational wave signal $h_\nu(f)$ is

$$h_\nu(f) = \frac{\nu(f)}{\nu} \frac{\Delta L_{opt}}{L_{opt}}, \tag{10.6}$$

where ΔL_{opt} is the difference in optical path length between the two arms. An unmodified argon ion laser just "out of the box" typically has frequency noise $\nu(f) \approx 10^4$ to 10^5 Hz/$\sqrt{\text{Hz}}$ for frequencies below 1 kHz.[135] Even with the stabilization techniques to be discussed below, a rather close balance between arm lengths is called for.

The lesson to be drawn from this example is that by building sufficient symmetry into the instrument, it can be made null with respect to a variety of perturbing influences. In this case, an equal arm interferometer compares the arm lengths by a clever subtraction that does not depend on the calibration of the "ruler", the laser wavelength. Symmetry has benefits against other noise sources as well. Below we'll see how excess noise (above shot noise) in the laser output power might corrupt the signal; symmetrical arm lengths help reduce this problem to a minimum as well.

10.1.5 *Active null instruments*

There is still another related use of the word "null" to describe good experimental practice. The concept of a *active null instrument* (also known as a "null servo") is an extension of some of the advantages of a null experiment to other experiments in which one does not necessarily expect a zero result. The key idea is to use another physical influence, distinct from the one being measured, in order to null out the reading that would have been caused by the object of measurement. The original measurement problem is replaced by the problem of measuring the size of the compensatory influence necessary to cause a null reading. Said this way, it almost sounds preposterous, but this is a technique that is often quite useful as a defense against a variety of experimental difficulties.

Again, the theory of this technique was propounded by Dicke.[136] He was considering the measurement technique known as *microwave radiometry*, the measurement of RF power. (Dicke spent World War II at the Radiation Laboratory at MIT working on the development of microwave radar.) The standard way to measure the *radiation temperature* of a radio source (the temperature of a perfect black body with the same radio brightness) was to aim a radio receiver (antenna, amplifier, and power-detecting diode) at it and to read the received RF power on a meter. Such a DC measurement is fraught with difficulties, not least of which is that the accuracy of the measurement depends on the calibration of the receiver. This is especially true if the radiation temperature of the source is small compared to the *noise temperature* T_n of the receiver, for then any dirft in the gain of the amplifier will multiply the large receiver noise, swamping the signal. The amplifiers in use at the time had big problems with gain drifts, so such measurements were hard to make precisely.

Dicke's insight was that sensitivity to drifting gain was much reduced if the original measurement was replaced by a measurement consisting of a comparison between the temperature T_{tar} of the radio source in question (the target) and a reference source of similar and known temperature T_{ref}. The receiver is made to switch rapidly between looking at the target and the reference. It is easy to see that if the reference source happened to match the target object's radiation temperature exactly, then one would always get the same receiver output, whether the receiver is aimed at the target or the reference, independent of the gain of the receiver. Appropriate circuitry can calculate the difference between the receiver's output when looking at the target and its output when aimed at the reference; when $T_{tar} = T_{ref}$, the output of this circuit will equal zero. A schematic diagram of such a *Dicke radiometer* is shown in Figure 10.2. In a very real sense, this instrument is

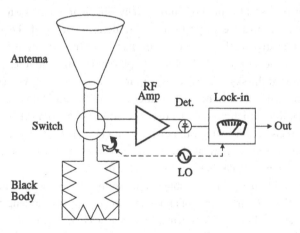

Figure 10.2: A schematic diagram of a Dicke radiometer, an example of a null instrument. If the temperature of the black body is adjusted to match that of the radiation entering via the antenna, then this would be an example of an active null instrument.

a microwave analog of the classic bridge circuit. The only major difference is that instead of making the comparison between two special points in the circuit as in a classic bridge, the subtraction is done by the switching technique, also known as *lock-in detection.*

A Dicke radiometer works because a zero difference still appears to be zero even if the receiver gain should drift by a substantial amount, as long as the drift is slow compared to rate of the switching between target and reference. Even if the match is not exact, sensitivity to gain drifts ΔG is still reduced. The effective error in temperature ΔT_{tar} due to gain drifts is

$$\frac{\Delta T_{tar}}{T_{tar}} \approx \frac{T_{tar} - T_{ref}}{T_{tar}} \frac{\Delta G}{G},\tag{10.7}$$

instead of $\Delta T_{tar}/T_{tar} \approx (T_{tar} + T_n)\Delta G/G$ for the naive receiver. The analogy to a bridge circuit's insensitivity to drive voltage is clear.

Of course, this scheme assumes that you know the strength of the reference source. In microwave engineering, this is often the case. It is pretty easy to make a good approximation to a black body at microwave frequencies and to measure its temperature. Thus one has a reference source of microwaves of known steady strength. To fully implement an active null radiometer, one would adjust the temperature of the reference source until the differential radiometer gave a null reading; temperature sensors in the reference source then give the radiation temperature of the target.

This was precisely the method employed by what is arguably the most beautiful microwave measurement ever performed, the determination of the spectrum of the Cosmic Background Radiation by the Far InfraRed Absolute Spectrometer (or FIRAS) on the COsmic Background Explorer (COBE) satellite.[137] One horn antenna pointed at the sky, while another input to the instrument was terminated in a carefully constructed black body. A further layer of "nullity" was provided by an external calibrator, which was placed over the input horn from time-to-time. The spectacular precision with which the instrument checked the thermal character of the CBR spectrum owes a great deal to the care with which the reference black bodies, and their associated thermometers, were designed.

The spectral information was obtained by applying the sky horn and reference black body to the two input ports of an interferometer whose arm length difference was rapidly varied. This clever scheme, known as a *Fourier transform spectrometer*, gives an output that is proportional to the Fourier transform of the spectrum of the incoming radiation. For a description of this style of instrument, the reader is referred to the specialized literature, such as for example the book by Bell.[138]

10.2 The Advantages of Chopping

There are further benefits to performing a *chopped* measurement, the generic name for a measurement made by rapidly switching between direct and reference measurement. Chopping adds a characteristic time- or frequency-domain signature to the instrument output, proportional in strength to the difference between the signal and the reference inputs. In the simple case we are discussing, the signature is a square-wave modulation, alternately on and off. Noise arising "downstream" from the chopping mechanism does not have the same signature applied, but is simply added in. By looking in the data stream for that particular signature, the signal can more easily be distinguished from noise.

Dicke can't have been the first to understand this concept, but he did recognize the power and general utility of the idea. Generalizing from his radiometer work, he invented the by now ubiquitous *lock-in amplifier*, a general purpose laboratory instrument to control the modulation of measurements and to *synchronously demodulate* the signal from the data stream. In optics labs, one often sees a many-slotted chopping wheel driven by a motor, used to modulate signals by periodically interrupting a light beam. They are almost always used together with a lock-in, which demodulates the signal of interest by flipping the sign of an internal amplifier in synchronism with the chopping, then averaging the output of that amplifier for many chopping cycles. In this way, an electrical signal is created whose magnitude is proportional to the difference in the sensor outputs in the "on" and "off" states.

What does chopping really do? The signal one wants to look at is modulated, so that its strength appears as the amplitude of a relatively high frequency sinusoid (and its harmonics, if one chops with square wave modulation.) If you average for many cycles (say, for a time τ_{int}), as is good practice, then the signal has to compete only with the noise that happens to match the chopping signature, that is in a narrow band of width $\sim 1/\tau_{int}$. This gives a noise advantage over quicker measurements, with their inherently larger effective noise bandwidths.

But how are we to understand the advantage of rapid chopping in these terms? We could always average a DC measurement for a long time, and thus obtain the same bandwidth narrowing effect. The noise advantage can only come about if there is a smaller spectral density of noise at the modulation frequency than near zero frequency, where unchopped measurements are made.

It is one of the facts of life, for which no general explanation exists, that low frequency measurements are almost always corrupted by excess noise. This is usually called $1/f$ *noise*, since low frequency noise power spectra usually vary like $1/f^{\gamma}$. (The index γ can actually vary over a wide range from one situation to the next; we'll

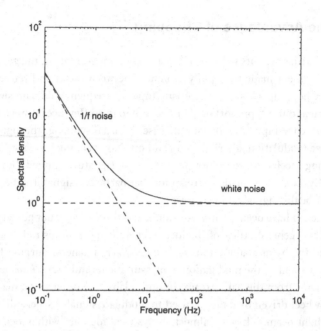

Figure 10.3:   An example of a noise spectrum exhibiting white noise at high frequencies, with a component of $1/f$ noise at low frequencies.

use the term $1/f$ noise whenever γ is greater than zero, although specialists often reserve the term to refer to cases when γ is closer to unity than to either 0 or 2.) A generic noise power spectrum is something like the one shown in Figure 10.3. It has a $1/f$ like region at low frequencies, disappearing into a white power spectrum above some corner frequency f_{cor}. It is clear that for power spectra like the one shown, one is much better off with a modulated measurement, fighting the low level of white noise, than with a DC measurement corrupted by $1/f$ noise.

A stimulating survey of a broad range of processes exhibiting $1/f$ noise was given by Press.[139] A more in-depth discussion of electrical $1/f$ noise can be found in the review by Weissman.[140]

10.3   The Necessity to Operate a Gravitational Wave Interferometer as an Active Null Instrument

The techniques of active null instrument design, including chopping, are extremely valuable in gravitational wave interferometers, even though it is impossible to modulate the flux of gravitational waves onto the instrument. The spectacular properties that would be required of the material from which one could construct a gravitational wave chopping wheel are described in another amusing article by Press.[141]

10.3.1 *The need to chop*

When we discussed optical readout noise in Chapter 5, we implicitly assumed that the noise in the light hitting the photodiode had the minimum possible noise power consistent with the quantum nature of light. But the power spectrum of the output power from a laser has a substantial $1/f$ component. For example, the argon ion lasers used to date in interferometers exhibit output power noise in excess of shot noise for all frequencies up to a corner frequency of several MHz or higher. This means that naive measurement strategies could not get close to the "shot noise limit" — an argon laser may show low frequency fractional power fluctuations of $10^{-5}/\sqrt{\text{Hz}}$,[142] several orders of magnitude greater than shot noise fluctuations in 1 Watt.

To see how this excess noise might corrupt the output, imagine that the interferometer masses were in positions that made the light at the output have a power level P_{out} that is half of maximum possible value. This was the "naive" operating point that we chose to discuss in Chapter 5; its chief advantage is that the interferometer yields the greatest change in output power for a given change in arm length difference when operated near that point. A shift in the optical path length difference between the two arms, whether due to a gravitational wave or to the noisy motion of the test masses, will cause a change in the output power. One direction of length difference change causes a power increase, the other sign a decrease. So, as we said in Chapter 5, the interferometer is a rather elaborate transducer from optical path length difference to output power.

In use, we have to assume the light has a constant power, and interpret any power change at the output as a shift in the light travel time difference, perhaps caused by a gravitational wave. Indeed, the fluctuations in power that we call shot noise cause noise in the measurement because we can't distinguish a power change caused by an arm-length shift from a power change caused by a random fluctuation in the photon arrival rate. (We'll discuss how one might test if a "blip" might really be a gravitational wave in Chapter 14.)

Low frequency excess intensity noise in the laser would be translated into noise in the gravitational wave measurement in the same way as shot noise. Thus, it is important to find a way to chop the signal, so that the path length difference information can at least be encoded in a frequency range limited by shot noise only.

10.3.2 *The need to actively null the output*

Before we can proceed toward inventing a proper chopping scheme, it is important to recognize that without something further, the operating procedure described above would not work at all. We assumed that the interferometer was undergoing small excursions about a particular operating point with a well-defined average output power and, more importantly, a well-defined derivative of output power with respect to optical phase difference $dP/d\Phi$. This would be a fine assumption, if all we had to worry about were the flux of gravitational waves, optical readout noise (including radiative pressure noise), and Brownian motion.

But the large magnitude of the low frequency seismic noise spectrum makes this naive interferometer design unworkable. To see why, let's make a numerical estimate of the path length difference caused by seismic noise. The rms seismic noise amplitude from a few tenths of a Hertz to 1 Hz is roughly $x_{seis} \approx 1$ μm. This is a band in which most of the vibration isolation systems we've described do not offer any isolation at all. Any high Q resonances, such as that of the high Q pendulum suspension for the test mass, multiply the rms motion in the vicinity of the resonance many-fold, specifically by $\sqrt{Q}$. Furthermore, the folding of the optical path translates a test mass motion x into a path length shift $\mathcal{N}x$, where $\mathcal{N}$ is the number of path round trips, or into $(2/\pi)\mathcal{F}x$ for a Fabry-Perot cavity with finesse $\mathcal{F}$. Finally, there are two arms, each with a mirror at each end; as the noise inputs at each should be statistically independent of the others, we multiply by a factor of $\sqrt{4}$, or 2. Thus we have a root mean square path length difference of order

$$\delta x_{rms} \sim 2\mathcal{N}\sqrt{Q}x_{seis}. \tag{10.8}$$

or, even if we could ignore the high Q enhancement, about 60 μm if we use $\mathcal{N} = 30$. If we use green light, such as from an argon ion laser, then a whole fringe shift occurs in a path length difference change of $\lambda = 0.5$ μm. So our naive interferometer would, at its rms noise amplitude, be swinging through many fringes. Trying to extract any gravitational wave information in the way described above would be very difficult.

(Ignoring the resonant enhancement factor $\sqrt{Q}$ will be justified by a feedback technique to be described in the next chapter. There we will see that a technique called *active damping* can cause the pendulum to act like a low Q oscillator near the resonance, yet still keep the low thermal noise associated with a high Q oscillator, at least at high frequencies.)

So the simple scheme won't work. But there is no need to despair. Operating the interferometer as an active null instrument will save the day. The key is to use *feedback* to keep the interferometer fixed at a chosen operating point, such as the

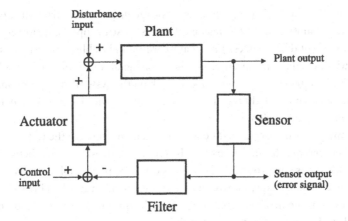

Figure 10.4: A block diagram showing the parts of a generic feedback control system.

mid-power point of a fringe. (Actually, we'll soon choose a much better operating point.)

In a gravitational wave interferometer, making a null detector solves a problem even more severe than the gain drifts Dicke solved with his null radiometer. In effect, an interferometer swinging through fringes has a gain $dP/d\Phi$ that varies between zero and maxima of both signs.

To implement this feedback scheme we need to add three subsystems to the interferometer. The first, a *sensor*, produces an *error signal* measuring how far the interferometer is from the desired operating point. The second applies an influence to the interferometer to bring it toward the desired operating point. The device that supplies the feedback influence is generically called an *actuator*. The input to the actuator comes from the error signal, usually filtered by the third key subsystem, a *compensation filter*. A schematic diagram is given as Figure 10.4.

We can then take as the output of the feedback interferometer, not the output power (since that will be held near a fixed level), but the strength of the feedback influence necessary to hold the system at the operating point. In other words, we have turned a gravitational wave interferometer into an active null instrument of the classic type.

There are many ways to implement each of these subsystems. Let's consider simple versions of each. To generate the error signal, we might use a photodetector at the output port of the interferometer, whose signal is applied to one of the input terminals of a differential amplifier. The other input of the amplifier is fixed at a level chosen to match what the photodetector's output would be at the desired operating point.

Since the source of the interferometer's large displacement from the operating point is actual motion of the test masses, it makes sense to apply some forces directly to one or more of those masses. Note that we wouldn't be able to make them sit truly "still" (whatever that may mean) using the error signal we've chosen. Instead, when the feedback system is operating properly, locking the system on an operating point serves merely to fix the light travel time difference at some chosen value, up to some integer number of wavelengths.

To apply the force, one might choose to attach magnets to the test masses, and adjust the currents through nearby coils. Or, an electrostatic force between high voltage electrodes and grounded test masses can be applied. Noise is a crucial issue here, as is dynamic range; clearly the actuation scheme must be able to compensate for the seismic driving term, or else it can't fulfill its function. Trade-offs are required, and are an interesting part of the design process.

It must be obvious that we couldn't even construct a feedback system without an actuator, and without a means for generating an error signal. The purpose of the compensation filter is more obscure, but it too is usually crucial. To understand why, we must take a brief look into the beautiful and rich theory of feedback control systems (also referred to as *servomechanisms* or "servos" for short.) That is the subject of the next chapter. Then, in Chapter 12, we'll return to the question of how to operate a gravitational wave interferometer as a null instrument, and to a quantitative discussion of the benefits that we obtain.

Feedback Control Systems 11

In this chapter, we give a sketch of the basics of the theory of feedback control systems. Our purpose is to give the reader enough background to understand the basic design principles of gravitational wave interferometers. To obtain a practical working knowledge of control system design, the reader is encouraged to study specialized texts; a good one at the beginner's level is Franklin, Powell, and Emami-Naeini's *Feedback Control of Dynamic Systems*.[66]

The subject of feedback has a rich history, with early contributions by Watt and Maxwell.[143] Military applications accelerated the development of this subject, like so many others, during World War II. Its crucial importance in aeronautics and spacecraft design has provided a continuing impetus for progress in the field during the post-War period.

First, a few words about terminology. The term *control system* refers to a machine whose output follows an input that we control at will. An example of great interest 50 years ago was a mechanism that caused a large gun to be pointed in the direction toward which an operator aimed a sight by pushing a lever. The term *servomechanism*, often shortened to *servo*, is also applied to systems of this sort; the name is derived from the word for "slave", referring to fact that the mechanism is designed to follow the command of the input while supplying much more power than is applied at the input. For a gravitational wave interferometer, we are interested in a special case of control system, one whose function is just to sit still at a chosen operating point; we can think of it as a control system with a null command input. This kind of system is sometimes called a *regulator*. Finally, the term *feedback* describes a strategy of control by which the system's actual output is compared with the desired output, the difference being used to adjust the system into the desired state.

Let's first consider a general feedback system, as shown in Figure 11.1. It has four generic parts. The system to be controlled, the generalized equivalent of the

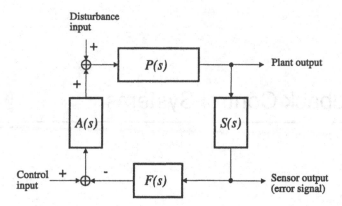

Figure 11.1:   Each component of a feedback control system can be characterized by its transfer function.

interferometer, is sometimes called the *plant* in servo jargon. (This usage comes from an industrial example; often one wants to use feedback techniques to control the operation of a large chemical process, say. In the more general context, we apply it to the parts of our systems whose design is determined by factors other than just suitability as a control system.) The system also includes a sensor to generate the error signal, a compensation filter and a feedback actuator. We will want to treat each of these parts as a linear time-invariant system such as we discussed in Chapter 4, with its output equal to the product (in the frequency domain) of the input times its gain.

We can call the gains of the four components $P(s)$, $S(s)$, $F(s)$, and $A(s)$ respectively. (It is useful to take as the frequency the complex variable $s = \sigma + i\omega$, since Laplace transform methods play a key role in understanding control systems.) The arrows in the figure indicate the direction of "signal flow." Note that while $F(s)$, the gain of the filter, is dimensionless, the other gains have dimensions associated with their functions as transducers between different sorts of physical quantities.

It may seem odd that, after introducing control systems as a solution to the dramatic non-linearity of the interferometer output with respect to optical phase difference, we immediately assume that all parts of the control system are linear. The strong temptation to do so comes from the fact that the most powerful tools for gaining insight into the behavior of a control system (and dynamic systems in general) only apply if the system is linear. Fortunately a working control system, by virtue of the fact that it keeps itself "locked" very near to a chosen operating point, has turned itself into a linear system; since it only makes small excursions from its operating point, we can define the gain of even a wildly non-linear system like an interferometer as the slope of the gain curve at the operating point. The assumption of linearity is justified after the fact.

Of course, the insight we glean will be inapplicable if the system is driven outside its linear dynamic range. We also learn little of the dynamics of the complicated process of "acquiring lock". For the latter, only direct integration of the equations of motion can supplement empirical understanding.

11.1 The Loop Transfer Function

We pay attention to the transfer functions of the individual parts of a feedback loop because we can learn about both the benefits and the costs of feedback from the study of them. Most important is the product of all of the gain factors in the system, known as the *loop transfer function* $G(s)$, given by

$$G(s) = P(s)S(s)F(s)A(s). \qquad (11.1)$$

This is the transfer function of the whole chain of components, with the feedback disconnected. Note that it doesn't matter whether we break the loop at the *summing junction* where the control input is applied or at any other point. We can imagine "cutting the wire" anywhere, taking the downstream end as the input to the system, and the upstream end as the output. Note also that, while P, S, and A have nontrivial physical dimensions, the loop transfer function $G(s)$ must be dimensionless, since it represents a set of transformations that restore a signal to the same form it had at the input — a signal must have the same dimensions at either end of a "wire".

One nice feature of the loop transfer function is that it can be measured directly, at least for real sinusoidal frequencies, for which $s = i2\pi f$. An old-fashioned yet heuristically suggestive way to measure it is to apply a sinusoidal signal of adjustable frequency to whatever point in the loop is convenient to take as an input. Display that input, as well as the output taken at the point that would have been connected to the input if the loop were closed, on a dual-trace oscilloscope. The magnitude ratio and relative phase of the two signals as a function of frequency can be read off the scope, and are a good way to represent $G(i2\pi f)$. (See Figure 11.2.) One often plots the magnitude and phase of $G(i2\pi f)$ in the form of a Bode plot, described in Chapter 4.

In a well-equipped modern lab, you'd replace the signal generator, dual trace scope, and hours of hand measurements with a two channel *signal analyzer* that digitizes input and output signals and computes the transfer function v_{out}/v_{in} via the Fast Fourier Transform (or FFT) algorithm. (This is a two channel version of

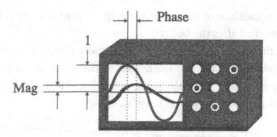

Figure 11.2: One method of measuring a transfer function, by comparing a system's output to its input using a dual-trace oscilloscope.

the spectrum analyzer mentioned in Chapter 4.) Don't try to build anything other than the simplest servo without one of these "FFT boxes".

With the loop open, the system might spend most of its time outside the linear range near the chosen operating point. This would make it difficult to measure the loop transfer function. If that is the case, it usually suffices to measure the gains of individual blocks separately, perhaps supplemented by a calculation of the gain of the crankiest element. Then, by definition, the loop transfer function can be found as the product of all the individually measured gains. This is another instance of the power of frequency domain methods — if we insisted on working in the time domain with impulse response functions, then we'd have to perform a set of convolution integrals to find the impulse response of the whole loop.

11.2 The Closed Loop Transfer Function

The loop transfer function $G(s)$ encodes the information about the *benefits of feedback*, as can be seen in an important theorem. Consider any servo system, and choose any two points as an input and an output. Call the transfer function between those two points $H(s)$ when the loop is *open*, that is when the feedback is disconnected. When the feedback loop is reconnected, or *closed*, the *closed loop transfer function* $H_{cl}(s)$ between the two points is equal to

$$H_{cl}(s) = \frac{H(s)}{1 + G(s)}. \tag{11.2}$$

To see why this is true, consider the system in Figure 11.3. We'll prove Eq. 11.2 for the important special case that the transfer function $H(s)$ represents the plant transfer function $P(s)$, but the argument can be easily generalized for the transfer function between any two points in the servo. First, consider the open loop situation, when there is no feedback. Then, the only input to the plant is the disturbance input

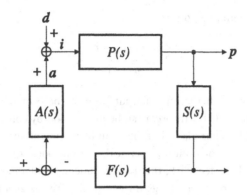

Figure 11.3: The signals necessary for understanding the closed loop transfer function of a feedback control system.

d. Thus, the plant output p will be $p = Pd$; that is, the open loop transfer function from the plant input to the plant output is just P.

Now, consider how the system behaves when we close the servo loop. One crucial change is that the input to the plant is no longer simply the disturbance input d, but is instead

$$i \equiv d + a, \tag{11.3}$$

where a is the feedback input applied by the control loop. The question is how this changes the transfer function of interest, the plant output p in response to a disturbance input d. The benefit of feedback comes from this additional signal a, whose value can be seen (by inspection of the figure) to be

$$a = -iPSFA \equiv -iG. \tag{11.4}$$

In other words, this additional input to the plant is proportional to the total plant input i, as it must be from the construction of the servo. This means we can rewrite Eq. 11.3 as $i = d - iG$, or more suggestively as

$$i = \frac{d}{1 + G}. \tag{11.5}$$

So what the loop has done is produce a net input to the plant that is smaller than the original input by a factor $(1 + G)$. Since the input is smaller, the output is too, $p = Pi = Pd/(1+G)$. Thus we can write the closed loop transfer function between disturbance input and plant output as

$$P_{cl} \equiv \frac{p}{d} = \frac{P}{1+G}. \tag{11.6}$$

The generalization to other input or output points in the system is straightforward.

The importance of this theorem can be understood by considering an unfed-back interferometer as our plant. Let the input be motion of one of the test masses. Take the output to be the optical power at the output port of the interferometer's beam splitter. Without feedback, the transfer function will almost certainly be so large that the interferometer swings through many fringes as a result of the low frequency component of seismic noise; this gives the interferometer an awkward, hard-to-read non-linear output. But with sufficient loop gain G, the transfer function can be reduced so much that the interferometer output stays very close to a chosen level of output power, exhibiting only small (and thus linear) output swings due to the seismic noise. Then, even if we choose to ignore these low frequency excursions as almost certainly due to terrestrial disturbances, higher frequency signals may still be read clearly superposed on the low frequency noise. Thus, the use of feedback to achieve a linear relationship between inputs and output gives a powerful benefit — we are able to use the fact that linear systems obey a superposition principle to read a signal at one frequency while ignoring noise at another.

At first glance, this result may seem like magic. How does the servo "know" how to produce just the right feedback input a, and how can it do it in time? If this bothers you, you are not the first. One way out of this confusion is to think in frequency domain terms, by considering as disturbance inputs sinusoidal signals of various frequencies. As long as the frequency is low enough, it should seem plausible that the servo can keep up. The question of what happens when the frequency is too high is an important and deep one. We'll return to it below, when we discuss the question of servo *instability*.

11.3   Designing the Loop Transfer Function

Often, the H in Eq. 11.2 represents a transfer function that one wants to minimize. A case in point would be the example we just discussed, the transfer function from seismic noise input to interferometer brightness output. Then, feedback can help if we make the loop transfer function G large. For all frequencies for which $G(s) \gg 1$, we have $H_{cl} \approx H/G$. For many such purposes, we'd like to make G as large as possible.

For some other applications, we're less interested in minimizing a transfer function H_{cl} than in "tailoring" its dynamic response. Feedback is also useful for this purpose. To see this, consider the function $1 + G(s)$, the denominator of the closed loop transfer function H_{cl}. The complex frequencies s at which $1 + G(s)$ vanishes, called the *poles* p_i of the system, determine the dominant features of the system's dynamic response, since H_{cl} blows up at those frequencies. The poles of the system are poles for all transfer functions. By adjusting the form of $G(s)$, we can place the poles of the closed-loop system where we wish.

The crucial theorem that applies here is the following: Any linear system whose dynamics are described by ordinary differential equations will have transfer functions that are rational functions of the frequency s, that is they can be written as a ratio of polynomials in s. As long as the system also obeys causality, then its transfer function must eventually *roll off* at high frequency, which implies that the degree of the denominator polynomial (call it n) must be greater than that of the numerator. Under these conditions, any transfer function of the system can be expressed as a *partial fraction expansion*, of the form

$$H_{cl}(s) = \sum_{i=1}^{m} \frac{a_i}{s - s_i} + \sum_{j=m+1}^{n} \frac{b_j s + c_j}{s^2 + q_j s + r_j}, \tag{11.7}$$

(or a variation of this form if a denominator polynomial has repeated roots.)[144] The s_i and the roots of the quadratic denominators in the second sum are the poles p_i of the system. The constants a_i, b_i, and c_j carry the specific information about gain from a particular input to a particular output. But, no matter where one chooses the input and output, the poles p_i are the same; these are the characteristic frequencies of the system's response to an impulsive disturbance.

The time domain behavior is described, as we saw in Chapter 4, by the impulse response; it is the inverse Laplace transform of Eq. 11.7. So the impulse response of the system will always take the form of a sum of n functions, each equal to a constant times $e^{p_i t}$. It is worthwhile developing a little intuition for the character these exponential functions take, as a function of the location of the poles in the complex plane. (See Figure 11.4.) Purely imaginary poles ($s_i = i2\pi f_i$) give sinusoidal contributions (of frequency f_i) to the impulse response. Purely real poles give real exponential responses $e^{s_i t}$. Poles with both real and imaginary parts correspond to damped sinusoids (i.e. sinusoids with exponential envelopes.) Poles with non-zero imaginary parts must always appear in pairs of complex conjugates, so as to be able to generate the equivalent of both $\sin 2\pi f_i t$ and $\cos 2\pi f_i t$.

The partial fraction expansion, which most students encounter only as an algebraic trick handy for evaluating indefinite integrals, here expresses a deep physical

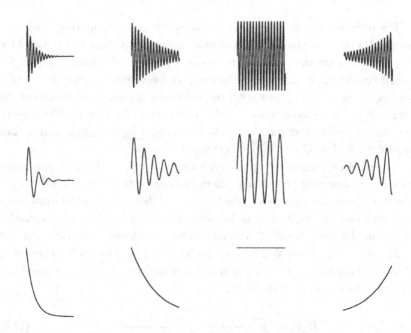

Figure 11.4: A palette of impulse responses of a linear system. They are arranged roughly according to the corresponding pole location in the s-plane. The imaginary axis is represented by the third column, the real axis by the bottom row.

truth. If you kick a linear system, its response will be a superposition of a characteristic set of behaviors that can be called "relaxation" (real poles) and "ringing" (poles with non-zero imaginary parts.) The functions $e^{p_i t}$ span virtually the entire repertoire of linear systems. The only exceptions occur when there are repeated poles; critical damping, with its two repeated real poles, brings in behavior like $t e^{p_i t}$. But the essentially exponential character of the solutions remains, as does the qualitative dependence of the solutions on the values of the poles p_i. Feedback can change the values of the p_i, but not the basic character of the solution.

11.4 Instability

In contrast to the benignly symmetrical roles of positive and negative imaginary parts, there is a crucial distinction between positive and negative real parts of poles. For, while the latter describe damped motion returning to an equilibrium position, positive exponents are associated with motion running away from equilibrium. In other words, poles with positive real parts describe unstable systems.

Instability is the ever-present worry of the control system designer. Of course, many ordinary (passive) systems exhibit unstable equilibria; consider a pencil balanced on its point. What is new with feedback is the possibility of turning a perfectly stable passive system into an unstable controlled system (if that isn't an oxymoron.) In the situations we are considering, an unstable system would be of hardly any use. The danger of instability is one of the most important *costs of feedback,* and avoiding instability is one of the most important constraints on control system performance.

11.4.1 *Causes of instability*

The reason instability is a ubiquitous possibility in feedback systems is that the benefits of feedback depend crucially on a particular sign choice for the feedback. At the summing junction, the influence from the actuator must counteract the other influences. Successful feedback is that which tends to drive the error signal to a null. Of course, it will never make a true null, just reduce its magnitude by a factor of $1 + G$. Feedback of the proper sign is often referred to as *negative feedback* for this reason.

Why can't we just choose the right sign for the feedback, then forget about it? The reason is that the sign of the feedback is not a simple binary choice. When a sinusoidal input, say $v_i \sin 2\pi f t$, is applied to a linear system, we expect a sinusoidal output, but there is no guarantee that the output will be simply proportional to $\sin 2\pi f t$, up to an overall sign. Instead, depending on the dynamics of the system the output may have an arbitrary *phase shift* with respect to the input; that is the output will have the form $v_o \sin (2\pi f t + \phi)$. The transfer function *phase* ϕ can take on any value; a shift of ϕ by $\pm 180°$ corresponds to a change in sign.

The phase of the feedback cannot be constant because we can only make a control system out of components of finite size, from materials obeying causality, that is in which causes precede effects. In mechanical systems, this is equivalent to requiring that there be a finite speed of sound. This is enough to guarantee that at high frequencies the output of a system cannot track the input, but must lag farther and farther behind. Usually, since our components have finite density as well, this effect is manifest as a series of resonances, each imparting an additional 180° phase lag.

As an example, consider a solid block of material of mass m and with, instead of infinite stiffness, a finite spring constant k. Assume that we are interested in controlling the displacement of the far end of the block x_{far} by moving near (input) end of the block x_{near}, under conditions in which there is no other load attached to the block. (This might be the case for, say, a mechanical indicator used as an

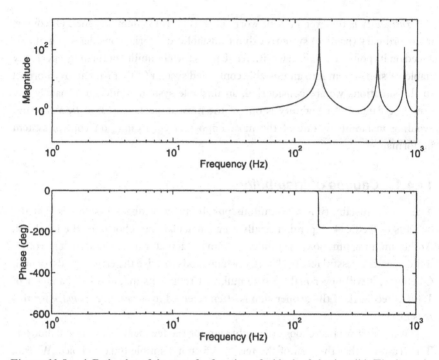

Figure 11.5: A Bode plot of the transfer function of a block of elastic solid. The input is the position of one end of the block; the output is the position of the opposite end.

output of some device.) Then it is not hard to show that the transfer function of this block is[145]

$$\frac{x_{far}}{x_{near}} = \frac{1}{\cos 2\pi \sqrt{m/kf}}. \tag{11.8}$$

This transfer function is graphed in Figure 11.5. The structure of such transfer functions depends on details of the boundary conditions, but the overall form, especially the succession of 180° phase shifts, is generic.

Thus even if the phase of a feedback system is set correctly at low frequency, it can't be correct to arbitrarily high frequency. This means that we can't reduce the requirements for stability to a simple statement of the correct feedback sign. A more subtle condition must be what distinguishes stable from unstable systems.

11.4.2 *Stability tests*

As we saw above, complex frequencies s where $1 + G(s) = 0$, or $G(s) = -1$, are special ones for understanding the behavior of control systems. It turns out, although it is less obvious than it seems, that purely imaginary (i.e. oscillatory) frequencies $s = i2\pi f_{unity}$ at which the magnitude of the loop transfer function $|G(i2\pi f_{unity})|$ equals unity are crucial for recognizing whether or not a servo will be stable. A useful rule of thumb (although not a precise statement of a theorem) is that a servo will be unstable if there is some *unity gain frequency* f_{unity} at which

(a) $|G(i2\pi f_{unity})|$ is falling below 1, and
(b) the phase of G lags $180°$ or more from the proper phase for negative feedback.

A rigorous test for stability based on the structure of $|G(i2\pi f)|$ is not hard to formulate, but we won't do it here. Instead, the reader is referred to the section on the *Nyquist test* in any good introductory book on control systems.

Consider a more commonly stated test for instability: that a servo is unstable if there is any frequency $i2\pi f_{180}$ at which (a) $|G(i2\pi f_{180})| > 1$ and (b) the phase of G lags $180°$ or more from the proper phase. Although the truth of this statement seems obvious from the considerations above, it is false. In fact, *conditionally stable* servos that fail this criterion turn out to be crucial for the operation of gravitational wave detectors.[146]

11.5 The Compensation Filter

Finally, we are in a position to examine the role of the compensation filter $F(s)$ in control system design. Consider the case in which the goal of the feedback system is to attenuate some external influence to the maximum degree possible. To achieve this, as we saw above in Eq. 11.2, we need to apply the largest possible $|G|$ at all interesting frequencies. But our ability to do this is limited by the requirement of stability. Mechanical resonances and other sources of phase lag in the loop transfer function cause the loop transfer function to eventually roll off in magnitude. Thus, we will only be able to achieve $|G(i2\pi f)| > 1$ in a restricted band, referred to as the servo's *active bandwidth*.

If we built a control system in which the compensation filter were just a constant gain independent of frequency, we would come closest to our goal by choosing the largest gain for which the system was still stable. The question is, "Can we achieve larger values of G within the active bandwidth, and still have a stable system, if we use a more complicated compensation filter than a constant gain?" Almost always, the answer is yes.

One common sort of compensation filter illustrates how we might do better. As we mentioned above, our stability problem is often due to resonances that change the phase of the feedback by $180°$ at high frequencies. But often, we are interested in achieving the benefits of large gain G mainly at low frequencies. A useful strategy is to roll off the loop gain, by using a low pass filter whose corner frequency is as low as possible. The overall loop gain is adjusted to make $G(f) < 1$ for all frequencies at and beyond the first resonance; then the servo will be stable. The frequency dependence of the gain ensures that even though G is small at high frequencies, it is large at low frequencies in order to obtain the benefits of feedback where we really need it.

Other uses of the compensation filter are to tailor the dynamic response by choosing where the poles of $G(s)$ will lie. For example, it is often desirable to avoid having poles p_i with very small real parts, as they correspond to systems that "ring" for a long time after being disturbed. Usually, we will need a more interesting $F(s)$ than a constant to achieve such optimum pole placement; a filter that shifts the phase of the feedback by $30°$ or so near the frequency of the offending pole usually does the trick. Even when our main goal is to maximize $|G|$ within the servo's bandwidth, it is usually desirable to compromise a little in order to place the system poles away from the imaginary axis.

Before proceeding to discuss those specific cases, it is worth considering why compensation filter design is an interesting (read "challenging") activity. For the job of maximizing gain, why not simply construct a compensation filter with an extremely steep slope in the graph of $|G(i2\pi f)|$ versus f? Or, why not fix the bandwidth problem entirely by constructing a filter that cancels the phase lag of the other components? To some extent, these goals can be achieved, but not with complete freedom. It is not possible to independently specify both the gain and phase of a transfer function as functions of frequency. If either one of these functions is completely specified, the other one is thereby completely determined. Alternatively, if either the real part or the imaginary part of $G(i2\pi f)$ is specified, the other is completely determined. This property follows from the requirement that any transfer function obey causality. The theorem embodying this relation is called the Kramers-Kronig relation[147] in the physics literature, and Bode's gain-phase relation among control engineers.

For an exact but rather opaque statement of this important theorem, the reader is referred to Franklin et al.'s book.[148] There is, however, a very simple approximation that you should always have at your fingertips. In frequency ranges in which the gain can be modelled as a power law $G \propto f^{-\gamma}$, the gain-phase relationship is approximately

$$\text{Phase}(G) \approx -\gamma \times 90°. \tag{11.9}$$

If an important requirement is that the servo's stability be robust, so that you can "set it and forget it", a single pole roll-off (often referred to as "6 dB/octave"), with a phase lag of 90°, is usually the best choice. For example, most operational amplifiers are compensated in this way. On the other hand, if one needs the very largest values of G that can be achieved, then a steeper roll-off will work, at the cost of needing more careful attention to the conditions of stability. The stability criterion suggests that a two pole roll-off, whose phase lag is 180°, is a good choice, but one needs to be careful when approaching the high frequency range in which the resonances occur. Filters that use a very steep roll-off at low frequencies, tailoring to a shallower roll-off at f_{unity}, are the answer for the most demanding applications. A detailed discussion of this aspect of control system design can be found in Horowitz's *Synthesis of Feedback Systems*.[149]

Bode's gain-phase relation imposes stern discipline on the activities of the designer. One's first idea about a compensation filter is often based on thinking only about either the gain or the phase. It is important to check whether the filter's effect on the neglected part of the transfer function tends to negate the intended effect. It often does. If so, don't despair; try again.

11.6   Active Damping: A Servo Design Example

As promised in Chapter 10, we can now consider how feedback can enable us to control the resonant response of a high Q pendulum. We'll see how we can achieve negligible amplification of input vibration at the resonant frequency, while still maintaining the low thermal noise associated with a low-loss oscillator.

First, let's consider how a servo can mimic passive damping. The key desirable feature of a dashpot attached to the mass in a mechanical oscillator is that it applies a force proportional to the velocity of the mass with respect to its environment. If we can invent a servo that also does the job, then we will have created something we can call *active damping*. There are many ways to actually implement such a system; here we consider one typical version. A schematic diagram of the elements of this servo system is shown in Figure 11.6.

First let's identify the standard parts of a servo: the plant, the sensor, the compensator, and the actuator. Here, the test mass on its pendulum suspension is the plant. One convenient kind of sensor consists of an LED illuminating an edge of the test mass, casting a shadow on a nearby photodiode. As the mass moves back and forth it modulates the amount of light falling on the photodiode, so the resulting photocurrent is a measure of the position of the mass with respect to the structure on which the sensor is mounted. Among the many possible choices of actuators, perhaps the most popular is to attach a small permanent magnet to the mass, and to place a coil in close proximity to it. When a current is driven through the coil, it applies a force to the magnet and thus to the mass.

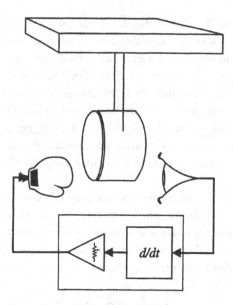

Figure 11.6:   A schematic diagram of a feedback system to damp the response of a pendulum.

How should a compensation filter be constructed for this choice of components? To mimic a dashpot, we'd like to have the control system apply a force that is proportional to the velocity of the mass. But we've proposed using a sensor that gives a signal proportional to the mass's position. (We could have decided to use magnetic pickup like that used in many seismometers. That would give a voltage proportional to velocity. But the photodiode sensor is very easy to use, and its position output is very handy for monitoring alignment.) If we choose as our compensator a circuit whose output is the time derivative of its input, then we have in effect converted a position sensor to a velocity sensor. The output of this differentiator circuit is applied to a driver amplifier for the actuator. For simple operation, the driver amplifier should be of the sort that produces an output current proportional to the input voltage[150]; then, there will be a fixed ratio of output force for a given input to the driver, even though the impedance of the coil depends on frequency.

Now that we've described all of the components of the loop, we can construct the loop transfer function $G = PSFA$. The pendulum transfer function P, defined as the amplitude of mass motion in response to a sinusoidal driving force of unit amplitude, is given by

$$P(f) = \frac{1}{k - (2\pi f)^2 m + ik\phi}, \tag{11.10}$$

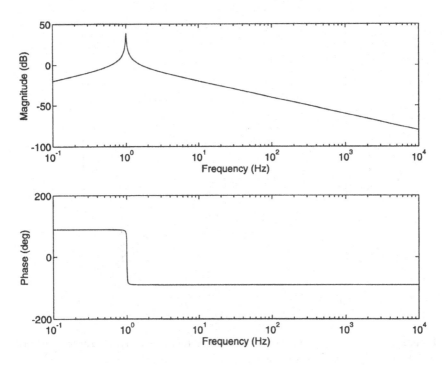

Figure 11.7:   A Bode diagram of the loop transfer function of an active damping system.

where we model the pendulum with an effective spring constant $k \approx mg/\ell$, and represent its damping with an appropriate $\phi(f)$. (Recall the discussion in Chapter 7.)

To a good approximation, $S(f)$ and $A(f)$ can be treated as constants, independent of frequency, if we want to consider the sensor as a position sensor and call the differentiator our compensation filter. An ideal differentiator has a transfer function $F(f) = i2\pi f$. The loop gain $G \propto PF$ is shown in Bode plot form in Figure 11.7. Note that the gain is sharply peaked at the plant's resonant frequency, falling to low values both toward high frequency and toward zero frequency.

If we were to hook up all of these components, then we'd be able to apply an amount of damping to the mass that is proportional to the gain of the active components. For convenience in tuning the system, it is a good idea to build in an amplifier with adjustable gain, as well as a simple way to flip the sign of the feedback (so you don't have to worry about getting all the signs right the first time), and a switch that can open or close the control loop.

The goal of all of this effort is to end up with a closed loop system that shows the same high-frequency response and noise spectrum as a passive high Q pendulum, but with a well-damped resonance instead. To see how this works, let's study the

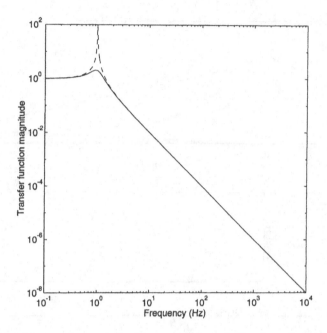

Figure 11.8:   The open loop and closed loop seismic transfer functions of a pendulum fitted with an active damping system.

transfer function $H = x/x_g$. In the open loop system, this will be, by Eq. 8.6,

$$H = \frac{f_0^2(1 + i\phi)}{f_0^2 - f^2 + if_0^2\phi}. \tag{11.11}$$

The closed loop version of this transfer function is, recalling Eq. 11.2, $H_{cl} = H/(1 + G)$. We would like to reduce the height of the resonant peak without too much effect at other frequencies. We can accomplish this by choosing an overall gain setting that makes $|G(f_0)| \gg 1$, but leaves $|G| \ll 1$ at frequencies far from the resonance. Happily, the form of G (Figure 11.7) allows us to do this easily. We show a Bode plot of H_{cl} for one particular gain choice in Figure 11.8.

It sounds like one could hardly go wrong. Is this true? Almost. One wrinkle is that the differentiator circuit will almost certainly oscillate unless we roll off its gain at high frequency. (This is yet another manifestation of the sort of bandwidth limitation any physical component must satisfy.) When we do that, with a capacitor in parallel with the feedback resistor, we no longer have a perfect velocity sensor. Does this matter? No. The performance of this servo is surprisingly insensitive to the fidelity with which the compensator produces a velocity signal. The ideal differentiator provides a gain proportional to f, and a phase lead (advance) of $90°$. But the servo damps the resonance almost as well if it has a phase lead of only $20°$ at the pendulum resonance, and correspondingly shallower increase of gain with frequency, per Eq. 11.9. Furthermore, neither the closed loop transfer function, nor even the stability of the system, depends much on what the filter does far from the resonance. In other words, this is a feedback system that is easy to make work, and hard to mess up.

But even if it is an easy job as control systems go, it still involves some trouble, some extra components, and the possibility of failure and even instability. It is important to understand the benefit one can achieve, now that the cost is apparent. In particular, what can be achieved through feedback that can't be achieved with ordinary passive damping?

We saw earlier, in Chapter 7, why applying passive damping to the test mass was unacceptable. The off-resonance power spectral density of thermal noise is directly proportional to the mechanical resistance at the mass, $\Re(Z)$. Only very low values of $\Re(Z)$ will yield a sufficiently low noise spectrum. Recall that we did not have to specify what physical mechanism is responsible for the damping in order to calculate the noise. To apply the fluctuation-dissipation theorem, we don't need to say whether the damping comes from gas molecule impacts, eddy currents, or an elasticity in the pendulum wire. All that is necessary to determine $x_{therm}(f)$ is the real part of the impedance.

What is magic about feedback damping? The crucial phrase in the statement of the fluctuation-dissipation theorem is that it holds for a system in thermal equilibrium. A feedback system is manifestly not a system in thermal equilibrium; unplug the amplifiers from the low entropy source of electrical energy that we use to power them, and they immediately cease to function. Thus since the fluctuation-dissipation theorem does not apply to a feedback damper, there is no prohibition on achieving a low noise level simultaneous with an appreciable degree of damping. This evasion of the fluctuation-dissipation theorem is the reason feedback damping is worth the trouble.

The reason this can work is that the mechanisms for damping are truly different in the active and passive cases. Passive damping comes about when a single degree of freedom in which we are interested is coupled to a bath of many other microscopic degrees of freedom. If the amount of energy in the interesting degree of freedom is greatly in excess of its equipartion value, then by virtue of the coupling energy becomes dispersed among the many degrees of freedom in the bath. But those other degrees of freedom act back on the one we care about through the same coupling mechanism. This is the reason for the link between dissipation and fluctuation.

Feedback damping works in a different way. With the control loop, one attempts to construct a device of only one other degree of freedom that nevertheless behaves as a more or less faithful reproduction of the damping process. Without the linkage between damping and the many degrees of freedom necessary for a passive isolator, there is no minimum required noise level for a given degree of damping. This is why a feedback system is allowed to evade what seems like a rather general law of physics.

Does this mean feedback damping is truly noiseless? No, not at all. The components of the loop are not ideal blocks in a diagram but real objects that perform their functions in a noisy way. There is another key theorem, due to Heffner,[151] that gives the minimum noise level for any linear amplifier, a category that includes the components of a feedback loop. To state the theorem, it is useful first to define the amplifier's *noise temperature* $T_{amp} \equiv v(f)i(f)/k_B$, where $v(f)$ and $i(f)$ are the amplitude spectral densities of the amplifier's input voltage noise and current noise, respectively. The theorem states that the minimum noise temperature that can be achieved by a linear amplifier is

$$T_{amp,\min} = \frac{1}{\ln 2} \frac{2\pi \hbar f}{k_B}. \tag{11.12}$$

The appearance of $\hbar$ indicates that this is a manifestation of the quantum mechanical uncertainty principle.

When we plug in actual noise spectra for particular sensors, we find that it is not especially easy to recover off-resonance noise spectra that are as low as the thermal noise in a high Q pendulum. This is especially true if we want to avoid the use of ultra-low noise sensors like interferometers, with their attendant complications. Assume we wanted to use a photodiode shadow sensor with a noise spectral density equivalent to mass motion of $x_n(f) \approx 10^{-11}$ m/$\sqrt{\text{Hz}}$, independent of frequency. This is roughly the shot noise in 1 mW of detected optical power, if the sensitivity is set so that the full dynamic range is about 1 mm.

What is the effect of this sensor noise on the output of the system? It is simplest to first consider the effect that sensor noise would have if the feedback loop were open. One way to picture this is to imagine that the shadow detector is moved away from the mass, so that no shadow falls on it. The noise in the sensor will cause a motion of the mass equal to

$$x(f) = Gx_n(f). \tag{11.13}$$

The block diagram in Figure 11.9 explains this: all of the elements of the feedback loop are involved in the transformation of noise at the input of the sensor (referred to equivalent motion at its input) into actual noise motion of the mass.

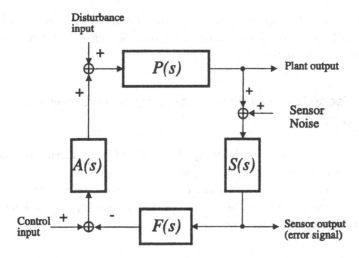

Figure 11.9: A block diagram of a generic feedback control system, showing how sensor noise enters.

The closed loop effect of this noise can, as before, be calculated from the rule $H_{cl} = H/(1 + G)$. Thus, the ratio of mass motion to sensor noise in the closed loop case is

$$\frac{x(f)}{x_n(f)} = \frac{G(f)}{1 + G(f)}. \tag{11.14}$$

This has two simple limiting cases. For frequencies at which $G \gg 1$, $x(f) \approx x_n(f)$. For the frequency range in which $G \ll 1$, we have instead $x(f) = G x_n(f)$, as in the open loop case.

For the damping servo, $G \gg 1$ only at the resonant peak. This means that the noise spectral density near the resonance must be at least $x_n(f_0)$. On the other hand, the high frequency limit of the loop transfer function is $G \approx f_0/f$. So, for $f \gg f_0$, the mass must move by at least $x(f) = x_n(f) f_0/f$. If $x_n(f) \approx 10^{-11}$ m/$\sqrt{\text{Hz}}$, this is not very good.

What can we do? We could, of course, imagine using a sensor comparable to the gravitational wave interferometer itself. But we needn't use such a gold–plated solution. If we are not too greedy, we might call the system successful if it achieved thermal noise limited performance above 100 Hz. Then, we can help our cause by using a filter in the electrical part of the feedback loop, rolling off the gain steeply for frequencies high compared to the resonance. Another useful strategy is to work not with a single pendulum but with a double pendulum, and apply the feedback forces to the upper mass only. This gives an additional $1/f^2$ filtering, here achieved mechanically instead of electrically. With careful application of such tricks, feedback damping can achieve its promise.

11.7 Feedback to Reduce Seismic Noise Over a Broad Band

Here, as further examples of the utility of feedback in gravitational wave interferometers, we give brief descriptions of two other techniques aimed at reducing the influence of seismic noise. Both will give isolation that works well near the pendulum resonance frequency. This could prove valuable, since purely passive isolators with very low resonant frequencies are hard to construct.

11.7.1 *Suspension point interferometer*

Feedback damping of the test mass suspensions is useful for minimizing the resonant amplification such suspensions would otherwise cause. The benefit of such systems is confined to reducing this extra noise in the vicinity of the pendulum resonance frequency. If we turned up the loop gain to try to get the servo to help us over a broader band of frequencies, we would only succeed in effectively locking the test masses to the surrounding structure, short-circuiting the vibration isolation system in the process. But if we instead monitor the seismic noise input,[152] we may in fact be able to reduce the effect of seismic noise on an interferometer over a broad band starting just above the pendulum resonance. A second interferometer, called a *suspension point interferometer*, can be installed to measure the difference in the separations of the tops of the test mass pendulum wires. When feedback is used to null this error signal, by a force applied to the top of one or more pendulums, differential length changes of the two arms are reduced. Since differential arm length changes are what a gravitational wave interferometer responds to, this scheme can effectively reduce the influence of seismic noise on the main interferometer output. It does this without reducing the latter's sensitivity to gravitational waves; they act directly on the main interferometer, irrespective of the action of this servo at the tops of the pendulums.

As beautiful as this idea is, it can't function perfectly. One reason is that for simplicity it will almost certainly be implemented as an interferometer with only a single round trip for the light. Since the main interferometer will be folded, noise from the beam-splitter enters at a reduced ratio; the suspension point interferometer won't properly account for this.

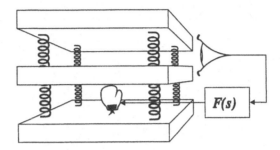

Figure 11.10: A schematic diagram of a passive vibration isolation system that has been equipped with a feedback system for active isolation.

11.7.2 *Active isolation*

Additional reduction of seismic noise motion of each individual test mass can be provided by a technique known as *active isolation*. An active isolation system for one degree of freedom starts with an inertial proof mass resonantly suspended from a movable platform. (See Figure 11.10.) A position sensor measures the displacement between the platform and the proof mass. The difference between the actual positions and the nominal DC spacing is the error signal that is amplified, filtered, and used to drive a mechanical actuator that can move the platform. The feedback, by nulling the error signal, causes the platform to track any motion of the proof mass.

Why does this provide vibration isolation? External vibration only moves the proof mass if, by moving the platform, the spring that connects the platform and proof mass is stretched — then it would apply to the proof mass a Hooke's Law force, $F = -k\delta x$. If the servo works to make the platform follow the proof mass, then that spring's length is being held approximately fixed, nulling the force that would have been applied to the proof mass. Furthermore, the platform is thus attached to the proof mass by an active "bootstrap" — since it is being servoed to follow the position of a well-isolated proof mass, it in turn responds less to external vibrations than in the open-loop state.

Single-axis active vibration isolation systems have been built by a number of groups, in both the geophysics and gravitational wave detection communities.[153] A group at the Joint Institute for Laboratory Astrophysics has built a six degree-of-freedom platform, for possible use in a second-generation LIGO interferometer.[154] Design challenges here include properly dealing with cross-coupling between the various degrees of freedom, and reduction of sensor noise to tolerable levels.

An Interferometer as an Active Null Instrument 12

Feedback is essential to the operation of a gravitational wave interferometer. Without it, an interferometer would not work - at the sensitivity required to detect astronomical gravitational waves, terrestrial disturbances would swamp the dynamic range of the instrument. Only by adding a precisely controlled countervailing influence, through a *fringe-locking loop*, can we hope to utilize the remarkable sensitivity of the interferometer. This is a classic application of the concept of an active null instrument. Once this is included in our picture of a gravitational wave interferometer, we will have sketched all of the basic ideas necessary to understand its successful operation.

We'll discuss two versions of a fringe-locking loop below. First, we investigate how to implement such a loop in a simple Michelson interferometer, or one involving a non-resonant folding scheme like the Herriott delay line. Then, we'll study the elegant version appropriate to locking a resonant Fabry-Perot cavity. Finally, we give an introduction to several refinements of the basic interferometer, in which feedback plays an essential role in enhancing interferometer performance.

12.1 Fringe-Lock in a Non-Resonant Interferometer

In Chapter 5 we discussed what we might call a "naive" interferometer. We chose as an operating point the maximum of $dP_{out}/d\Phi$, in order to maximize the sensitivity of the instrument. The most naive feature of the design was our assumption that the interferometer might stay in the vicinity of the operating point without feedback. We've been systematically demolishing that assumption ever since we first proposed it. In this section, we'll finally see how to use feedback to keep an interferometer near a chosen operating point.

While we're at it, we ought to reconsider whether the naive choice of operating point was the best one. While its advantages are obvious, it has an important drawback: since we use the output power as a measure of the phase, we aren't able to distinguish between a change in the laser's power and a change in Φ corresponding to a gravitational wave. (The output power change is proportional to the product of the input power and the phase shift.) Thus, if the laser has substantial excess output power noise in the gravitational wave signal band, as most lasers do, then our proposed scheme is highly vulnerable.

In the context of our subsequent discussion, especially that of Chapter 10, we can see that the naive scheme violates several of the general principles of good instrument design. Firstly, it works at DC, that is the signal appears at the output of the interferometer at the same low frequency as the gravitational wave's perturbation to space–time. There, in the audio frequency band, laser excess output power noise is high. (Recall the generic noise power spectrum of Figure 10.3; it applies to laser output power noise too.) Secondly, in an important sense, the instrument is not "null". In order to measure a small phase shift away from the operating point, we need to look for a small change in a large amount of optical power. It would be better if instead the power in the zero signal state were zero.

There is, of course, a choice of operating point that gives zero output power — where the phase difference is set so that the output port of the interferometer is dark. This is called, in the jargon of the field, the *dark fringe*. Our basic principles of instrument design argue for this choice of operating point. But our detailed knowledge of this particular instrument immediately shows a major difficulty. At the dark fringe, $dP_{out}/d\Phi = 0$. This must be true, since the power is a minimum there. Thus, if we try to reduce our vulnerability to one excess noise source, it seems we would be forced to set the gravitational wave sensitivity of the instrument to zero.

Or would we? Is there a way to meet the need for a chopped, null instrument, in the context of a feedback system that can actually keep the interferometer working at some particular operating point? Fortunately, there is. This crucial idea is due to Weiss.[17]

A scheme to achieve these goals begins with components like those shown in Figure 12.1. The new parts, one placed in each arm, give a way of modulating the phase of the light in each arm. This function is performed by an electro-optic device called a *Pockels cell*, placed just "downstream" of the beam splitter in each arm. A Pockels cell adds a phase shift to the light passing through it that is proportional to a voltage applied across it. This magic may be performed by crystals of lithium niobate ($LiNbO_3$) and potassium dihydrogen phosphate (KH_2PO_4, abbreviated KDP), among others, in which the dielectric tensor has a dependence on an external electric field. For more details, see for example Haus's Chapter 12.[155]

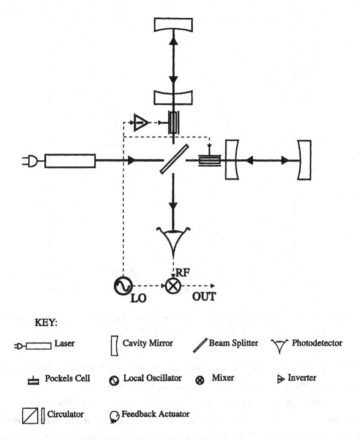

Figure 12.1: A schematic diagram of a Michelson interferometer equipped with Pockels cell phase modulators in each arm.

The phase shift available from a typical Pockels cell is not huge, only about one radian. But the speed of response can be quite high — bandwidths of many MHz are typical. This is crucial, since a modulation frequency of at least several MHz is typically required if the *interrogation* of the interferometer fringe is to be carried out at a frequency high enough that the power of the light source has no excess noise above the shot noise.

This interferometer fringe interrogation system is built around the choice of the dark fringe as operating point. Assume for now that the interferometer state is located close to the darkest part of the fringe, as illustrated in Figure 12.2. (We'll explain soon how it is kept there.) Now imagine that a sinusoidal voltage of some high frequency f_{mod} is applied in antiphase to the two Pockels cells. That adds a phase $\frac{1}{2}\delta \sin 2\pi f_{mod} t$ to the light entering one arm, while simultaneously subtracting

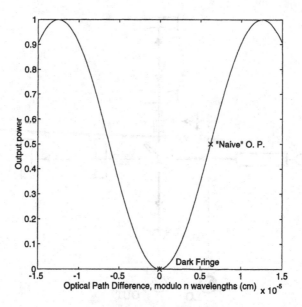

Figure 12.2: The output power of an interferometer as a function of path length difference, showing the locations of the "naive" operating point and the dark fringe.

$\frac{1}{2}\delta \sin 2\pi f_{mod}t$ from the phase of the light entering the other arm. The anti-symmetry is another good practice from the school of null instruments.

If the interferometer were centered on the dark fringe, then application of the modulation voltage causes the phase to oscillate about the minimum power point by an amount $\delta \sin 2\pi f_{mod}t$. How does that affect the power at the output port? The power exiting the output port will be given by (recall Eq. 2.12) $P_{out} = (P_{in}/2)(1 + \cos 2\Phi)$. The dark fringe, where $P_{out} = 0$, is centered on $\Phi_0 = \pi/2$. Thus, in the presence of a signal Φ_{sig} and the applied phase modulation, we have

$$P_{out} = \frac{1}{2}P_{in}(1 + \cos 2(\Phi_0 + \Phi_{sig} + \delta \sin 2\pi f_{mod}t)). \tag{12.1}$$

A useful expansion of the phase-modulated cosine term in Eq. 12.1 is known to students of special functions as well as to radio engineers[156]; it involves the Bessel functions of the first kind, $\mathcal{J}_n$. (See Figure 12.3.) We find

$$\begin{aligned}
P_{out} = (1 &- \mathcal{J}_0(2\delta)\cos 2\Phi_{sig} \\
&+ 2\mathcal{J}_1(2\delta)\sin 2\Phi_{sig}\sin 2\pi f_{mod}t \\
&- 2\mathcal{J}_2(2\delta)\cos 2\Phi \cos 2\pi(2f_{mod})t \\
&+ \cdots)
\end{aligned} \tag{12.2}$$

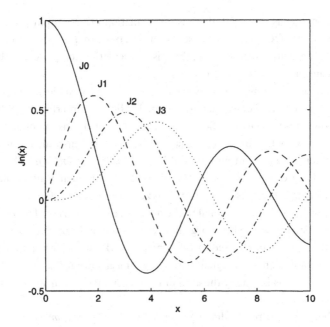

Figure 12.3: The Bessel functions of order 0 through 3.

This expansion is less frightening than it looks, since we are mainly interested in the case $\delta \ll 1$ radian. Then the Bessel functions have the simple approximations

$$\mathcal{J}_0(2\delta) \approx 1 - \delta^2, \tag{12.3}$$

$$\mathcal{J}_1(2\delta) \approx \delta, \tag{12.4}$$

and

$$\mathcal{J}_2(2\delta) \approx \delta^2/2, \tag{12.5}$$

with the higher order terms negligible.

Let's rewrite Eq. 12.2, making use of these approximations, and also replacing the trigonometric functions of Φ_{sig} with approximations good to first order. We find

$$P_{out} \approx \frac{1}{2}P_{in}(\delta^2 + 4\delta\Phi_{sig}\sin 2\pi f_{mod}t - \delta^2 \cos 2\pi(2f_{mod})t + \cdots). \tag{12.6}$$

When $\Phi_{sig} = 0$, then we have

$$P_{out} \approx \frac{1}{2}P_{in}(\delta^2 - \delta^2 \cos 2\pi(2f_{mod})t). \tag{12.7}$$

Note that, as long as $\Phi_{sig} = 0$, there is no oscillation of the output power at f_{mod}. What survives is a DC term and an oscillation at a frequency of $2f_{mod}$, both of order δ^2 in amplitude. But, when $\Phi_{sig} \neq 0$, there is an oscillating component of the power at f_{mod}; its amplitude is proportional to Φ_{sig}.

The foregoing discussion shows that this scheme has achieved its two goals, that is to construct a null chopped output signal. The naive interferometer had as its output a power proportional to the product $P_{in} \times (\Phi_{sig} + \text{a constant of order unity})$, varying at the gravitational wave signal frequency. Here instead, the modulated interferometer has as its output a power $P_{out} \propto P_{in}\Phi_{sig} \sin 2\pi f_{mod}t$. The signal of interest has been encoded as the amplitude of a high frequency sinusoidal power variation. Furthermore, this is a signal that is zero when the interferometer sits at its operating point, and is proportional to the departure Φ_{sig}. This is a far cry from the naive interferometer, which has a large non-zero output at its operating point. Note that even though our modulated interferometer works at the zero power point, it treats the algebraic sign of the signal correctly, giving an output $P_{out} \propto -\sin 2\pi f_{mod}t$ when $\Phi_{sig} < 0$. This is exactly the kind of interferometer output we were trying to construct.

Another concept from radio engineering, that of *modulation sidebands*, is very useful in describing how the signal has been encoded. The output at f_{mod} is proportional to $\Phi_{sig} \sin 2\pi f_{mod}t$, so when the signal is a function of time we have an *amplitude modulated* (or AM) signal

$$P_{out}(t) \propto \Phi_{sig}(t) \sin 2\pi f_{mod}t. \qquad (12.8)$$

For specificity, imagine that $\Phi_{sig}(t) = \Phi_0 \sin 2\pi f_{sig}t$. (We can always Fourier analyze an arbitrary signal into a set of sines and cosines.) Applying the appropriate trig identities, we can expand this expression to find

$$P_{out}(t) \propto \Phi_0(\cos 2\pi(f_{mod} - f_{sig})t - \cos 2\pi(f_{mod} + f_{sig})t). \qquad (12.9)$$

Note that, strictly speaking, there is no signal at precisely f_{mod} (unless $f_{sig} = 0$.) Instead, each signal frequency appears as a symmetrically placed pair of sidebands offset from the carrier frequency f_{mod} by the signal frequency f_{sig}. The technical term for this form of encoding is *double-sideband suppressed-carrier amplitude modulation*.[157] (The "suppressed carrier" part refers to the desirable feature that we have no term in the power proportional to $\sin 2\pi f_{mod}t$ when the signal is zero.)

To extract the interferometer signal Φ_{sig}, apply a voltage proportional to this photocurrent to the input of an *RF mixer*. This is a non-linear device, consisting of one, two, or four diodes, with two inputs for RF signals and one output. At the second input is applied the *local oscillator* signal $\sin 2\pi f_{mod} t$ used to drive the Pockels cell phase modulators. The output of the mixer is a voltage proportional to the product of the two RF input signals. This means the mixer's output is

$$V_{out} \propto \Phi_0 \sin 2\pi f_{mod} t (\cos 2\pi (f_{mod} - f_{sig})t - \cos 2\pi (f_{mod} + f_{sig})t), \quad (12.10)$$

or, using trig identities to expand the products of sinusoids,

$$V_{out} \propto \Phi_0 \sin 2\pi f_{sig} t + \frac{1}{2} \sin 2\pi (2f_{mod} - f_{sig})t - \frac{1}{2} \sin 2\pi (2f_{mod} + f_{sig})t. \quad (12.11)$$

The sidebands on either side of f_{mod} have been translated back down to their original frequencies. In other words, we have recovered a voltage proportional to Φ_{sig}, just as we wanted. There are also components near $2f_{mod}$. These are uninteresting here, and are easy to filter out.

This technique is called *heterodyne detection*, the same technique used in good AM radio sets. Note that it is also the radio frequency version of lock-in or phase-sensitive detection, as described above in Chapter 10. This is an example of the esssential unity of experimental physics — the same good idea is useful in many contexts.

12.2   Shot Noise in a Modulated Interferometer

It remains for us to inquire whether this scheme achieves its ultimate end, that of preserving the fundamental optical readout noise (shot noise) sensitivity of the naive interferometer, while removing the latter's sensitivity to excess output power noise from the laser. First, we calculate the shot noise in the modulated system.

The photodiode at the output port produces a current

$$I = \frac{\eta e \lambda}{2\pi \hbar c} P_{out}, \quad (12.12)$$

where e is the electronic charge, and λ is the wavelength of the laser light. The parameter η is the dimensionless *quantum efficiency* of the diode, denoting the fraction of incident photons that produce conduction electrons. It can usually be made close to unity. It is the "granularity" of this photocurrent, because it is made up of electrons of finite charge, that can be said here (in the semi-classical sense) to give rise to shot noise.

The shot noise in the photocurrent is given by

$$i(f) = \sqrt{2e\bar{I}}, \tag{12.13}$$

where $\bar{I}$ is the mean value of the photocurrent

$$\bar{I} \approx \frac{\eta e \lambda}{2\pi \hbar c} \frac{P_{in}}{2} (1 - \mathcal{J}_0(2\delta)) \propto \frac{1}{2} P_{in} \delta^2. \tag{12.14}$$

Recall that $i(f)$ is white, so $\sqrt{2e\bar{I}}$ is as appropriate at f_{mod} as it is at DC.

To see what this means in terms of gravitational wave sensitivity, we need to calibrate it in terms of something meaningful, such as Φ. We have

$$\Phi(f) = \frac{\pi \hbar c}{\eta e \lambda (P_{in}/2) \mathcal{J}_1(2\delta)} i(f). \tag{12.15}$$

Since $h(f) = \Phi(f)\lambda/2\pi L_{opt}$, this gives

$$h(f) = \frac{1}{2\pi L_{opt}} \frac{\pi \hbar c}{\eta e (P_{in}/2) \mathcal{J}_1(2\delta)} \sqrt{2} i(f_{mod} \pm f), \tag{12.16}$$

where we have displayed explicitly the fact that our system of modulation and demodulation translates the noise in both sidebands f away from f_{mod}, where the laser power is shot noise limited, down to the signal band f. Since the noise in the two sidebands adds incoherently, we pick up the extra factor of $\sqrt{2}$. Plugging in the expression for the shot noise, we finally find

$$h(f) = \frac{1}{L_{opt}} \sqrt{\frac{\hbar c \lambda}{\pi \eta P_{in}}} \frac{\sqrt{1 - \mathcal{J}_0(2\delta)}}{\mathcal{J}_1(2\delta)}. \tag{12.17}$$

This expression for the shot noise depends on δ, the amplitude of the phase modulation in the interferometer. But since we will only use $\delta \ll 1$, we can use the approximations of Eqs. 12.3 and 12.4. As $\delta \to 0$, the ratio $(\sqrt{1 - \mathcal{J}_0})/\mathcal{J}_1 \to 1$. This means that this modulation scheme has a comparable shot noise limit to the one we found for the naive interferometer.

(We oversimplified matters a bit in Eqs. 12.13 and 12.14, when we claimed that the shot noise in the photocurrent will be just that of a constant current equal to the mean photocurrent. The modulation of the interferometer phase makes the photocurrent vary; this means that the shot noise actually varies through the phase modulation cycle. This has the effect of increasing the shot noise somewhat above the simplified calculation. See the paper by Niebauer et al. for details.[158] The effect can be of order a 20% increment to the shot noise amplitude spectral density.)

12.3 Rejection of Laser Output Power Noise

Now let's consider how this interrogation scheme responds to fluctuations in laser output power that occur with frequencies within the audio frequency signal band. Refer to Figure 12.4. It shows the interferometer fringe pattern for two different values of the laser power P_{in}. The gain of the fringe phase readout (conversion of Φ_{sig} to light power modulation at f_{mod}) is proportional to the product $\delta\Phi_{sig}P_{in}$. Imagine Φ_{sig} to be fixed at some non-zero value and let P_{in} fluctuate. Then the amplitude of the photodiode output at f_{mod} will fluctuate, in a way indistinguishable from the case with fixed P_{in} and varying Φ_{sig}.

This is quite like the situation considered by Dicke in his classic paper on remedying gain variation in a microwave receiver, as we discussed in Chapter 10 above. Here, as there, the problem can be made to vanish if the measurement is carried out as a null measurement, since at $\Phi_{sig} = 0$ there is, to first order, no sensitivity to P_{in}. This, as we've argued already, is the reason that the dark fringe is the best choice for an operating point.

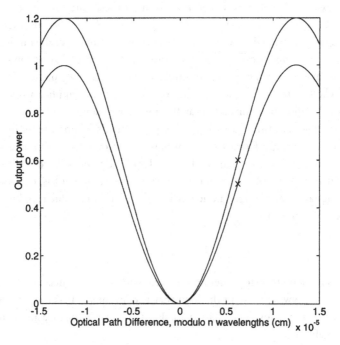

Figure 12.4: An illustration of how laser power fluctuations would cause interferometer output noise, if the system does not sit at the dark fringe.

12.4 Locking the Fringe

In order to realize this benefit, Φ_{sig} needs to be quite close to zero. External vibrations, if not counteracted, would move Φ_{sig} many radians. To make the interferometer work at all, feedback is required to keep $\Phi_{sig} \leq 1$ radian. To make it work well, the feedback needs to be effective enough to keep $\Phi_{sig} \ll 1$ radian.

How can this feedback be carried out? We've already described a system with all of the required components. Our plant is the interferometer as a whole. The sensor is the heterodyne phase detection system described above. The Pockels cells that apply the RF phase modulation can also apply low frequency phase shifts to the light entering the two arms. This means they can serve as the feedback actuator. Linked together in a loop, with the proper compensation filter, the system can function to hold the output state of the interferometer near the ideal $\Phi_{sig} = 0$. A schematic diagram of the loop is shown in Figure 12.5.

It is important to consider carefully how such a null instrument works, otherwise it might seem as if we are invoking magic. After all, aren't we using feedback to make the signal vanish? And if so, how can we measure anything at all?

One explanation that is sometimes given to these questions is to say that, although the sensor signal may be nulled, the feedback supplied by the actuator is tracking the true signal very accurately in order to accomplish the nulling. So, in effect, the measurement can be read out at the actuator instead of at the sensor. In the case we are discussing, we cannot directly read out the DC phase shift applied by the Pockels cells. But we can measure the voltage that is applied to the Pockels cell, only one calibration constant away from what we want.

There is some utility to this picture, but it tends to obscure as much as it illuminates. After all, a feedback system is built out of a set of (hopefully) linear elements cascaded together with well-defined gain. The Pockels cells wouldn't be correctly driven if the sensor output were truly zero. The point is, of course, that the error signal is never exactly nulled in a servo. Instead, its value is reduced below its open loop level V, to

$$V_{cl} = \frac{V}{1 + G}, \tag{12.18}$$

where G is the loop transfer function. So, we could choose to measure the signal at the sensor, or anywhere else in the loop that is convenient. It is only necessary to correct by the frequency dependent factor $1 + G$ to rescale the measurement to the open loop calibration.

This explanation perhaps still raises the question of what happens to the signal-to-noise ratio of the error signal. After all, if the signal amplitude is reduced by a

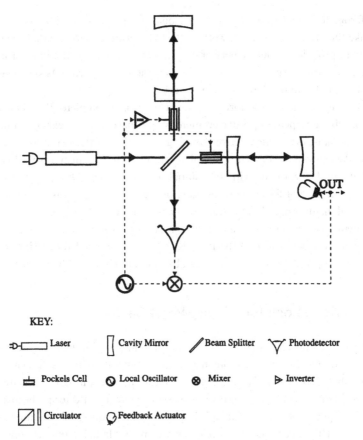

KEY:

Laser	Cavity Mirror	Beam Splitter	Photodetector
Pockels Cell	Local Oscillator	Mixer	Inverter
Circulator	Feedback Actuator		

Figure 12.5: A schematic diagram of a Michelson interferometer equipped with a fringe-locking servo.

factor of $1 + G$, doesn't that reduce the SNR by a like amount? Fortunately, the answer is no. To see why, consider noise at the input to the sensor. It causes an output just as if it were a true signal. This means that the feedback "corrects" for the noise in the same way as for a signal, and reduces its magnitude by the same factor. Thus, the fundamental signal-to-noise ratio of a closed loop system is the same as that of the system run open loop.

This assumes that the feedback loop obeys the standard design rule for multi-stage amplifiers: that no later stage should be so noisy that the noise it generates dominates the amplified noise from the *front end*, or first stage of amplification. Note also that the equality of the SNR of closed- and open-loop systems is only true for noise we might attribute to the sensor itself. The whole point of the feedback in the interferometer is to counteract the deleterious effects of external noise that threaten to make the interferometer not work at all. Furthermore, here the high servo gain is used to provide the nearly perfect dark fringe that reduces the interferometer's vulnerability to laser output power noise; so we actually improve its signal-to-noise ratio attributable to this latter source of noise.

One key feature of the foregoing description is incomplete. We calculated in Eq. 10.8 that an open loop interferometer might have a fringe excursion of many tens of radians, even when active damping is applied to the pendulums. But the typical Pockels cell can only supply a phase shift of order a radian or so. This means we need another actuator with substantial dynamic range. One possibility is the direct application of force to one or more of the test masses. This can provide the requisite dynamic range, but at rather low bandwidth. Use of both actuation schemes in tandem gives us, in effect, an actuator system with the dynamic range of mass forcing and the bandwidth of the Pockels cells. A more detailed description of the combined loop and its compensation can be found in Dewey's Ph. D. thesis.[159]

12.5   Fringe Lock for a Fabry-Perot Cavity

If we need a feedback system to lock an interferometer with delay-line arms, then we need it doubly for an interferometer with Fabry-Perot cavities as the arms. Recall that a Fabry-Perot cavity is an interferometer all by itself, in that the state of the light exiting it depends on the relative phase of the many interfering beams inside it. The response of the output light's phase to the motion of the mirrors is as steep for a Fabry-Perot cavity as it is for a comparable delay line, as long as the cavity is operating near resonance. But, without feedback, it is impossible to expect the cavity to stay near resonance, for exactly the same reason that a non-resonant interferometer won't stay near an operating point — the magnitude of low frequency seismic noise is too great. There is one additional tricky feature of a Fabry-Perot cavity: once it leaves the vicinity of the resonance, the state of the light leaving it has almost no dependence on the cavity length (until the next resonance.) This is because there is a steep dependence of total phase on cavity length only quite close to the resonance. (Recall the discussion of Eq. 6.18.) Colloquially speaking, it has large "dead" regions separating the resonances. This is unimportant to performance as long as the feedback loop is locked, but it does add an extra complication to the initial *lock acquisition*.

The extra complications are well worth overcoming. We discussed in Chapter 6 some of the reasons one might prefer to use Fabry-Perot cavities as the path-folding elements in long interferometer arms. In addition to that function, locked

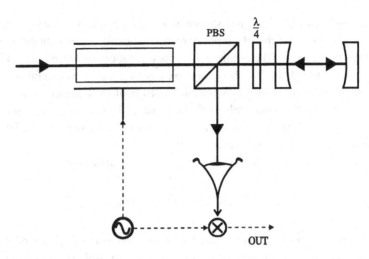

Figure 12.6:   A schematic diagram of the components used in the reflection cavity locking scheme.

Fabry-Perot cavities find several other key uses in gravitational wave interferometers. Frequency stabilization of the laser, desirable even in a non–resonant interferometer, is achieved by locking the light to a Fabry-Perot cavity used as a quiet length standard. And the technique of recycling that we'll discuss in the following section involves turning the entire interferometer into a single locked resonator.

For the moment, let's ignore which of these applications we have in mind, and consider a simple cavity of the type we examined in Chapter 6. Figure 12.6 shows the configuration of components used to implement the *reflection cavity locking* scheme.[160] It is an elegant technique, with roots in a related microwave technique (again!) invented by Robert Pound in 1946.[161] We'll also note its analogies with the locking scheme for non–resonant interferometers described in the previous section.

As before, we need a Pockels cell to apply phase modulation to the light entering the cavity. The other components in Figure 12.6, a polarizing beam splitter and a quarter wave plate, function as an optical *circulator*. The components are oriented so that the input light has the proper linear polarization to be transmitted by the beam splitter, but is then converted to circular polarization by the quarter wave plate. Upon reflection by the Fabry-Perot cavity, the quarter wave plate reconverts the circularly polarized light to linearly polarized light, only now it is of the orthogonal polarization to its initial state. This means it is reflected by the polarization sensitive beam splitter. This enables the returning light to be directed to a photodetector without blocking the path of the input light to the cavity.

Once again, a high frequency sinusoidal voltage is applied to the Pockels cell, so that the light that reaches the input mirror of the cavity is phase modulated. As we learned in Chapter 6, the response of a Fabry-Perot cavity depends strongly on the frequency of the light. For this reason, we'll need to pay more careful attention to the frequency content of this phase modulated light than was necessary in the non-resonant case. We avoid mathematical complexity by restricting the discussion to the case in which the *modulation index* δ is much smaller than one radian, the appropriate limit for this application.

We can write the electric field E_{in} in the phase modulated light as

$$E_{in} = E_0 \cos[2\pi f_c t + \delta \cos 2\pi f_{mod} t], \tag{12.19}$$

with f_c the frequency of the unmodulated light (the *carrier*) and f_{mod} the modulation frequency. As long as the phase modulation $\delta \ll 1$, we can avoid the Bessel function expansion and instead approximate Eq. 12.19 as

$$E_{in} = E_0 \cos 2\pi f_c t - \delta E_0 \cos 2\pi f_{mod} t \sin 2\pi f_c t. \tag{12.20}$$

This can be rewritten as

$$E_{in} = E_0 \cos 2\pi f_c t - \frac{\delta E_0}{2} \sin 2\pi (f_c + f_{mod}) t - \frac{\delta E_0}{2} \sin 2\pi (f_c - f_{mod}) t. \tag{12.21}$$

That is, the modulated light can be thought of as light at the carrier frequency plus light at two sideband frequencies, symmetrically spaced by f_{mod} on either side of the carrier. The phases of the sidebands (i.e. the appearance of sines where there once were only cosines) are crucial to understanding what is going on.

For a locking system, f_{mod} is chosen to be large with respect to the bandwidth of the cavity, Δf. The sensible choice for the operating point for this system is to make f_c equal to a cavity resonance frequency. These specifications are enough to determine how we can interrogate the relative match of the carrier with the cavity length. Recall how the phase of the reflected light depends on the tuning (as shown in Figure 6.5.) Far above or below the resonance, the reflected light has virtually no phase shift with respect to the incident light, but the phase shift goes through a full 360° as the frequency is tuned through the resonance. Exactly on resonance, the light suffers a 180° phase shift, in other words a change in sign.

Consider what happens to the carrier and two sidebands upon reflection. If the carrier is centered on resonance, its sign is flipped, while the sidebands, far from resonance, keep the same phase. The reflected light thus has the form

$$E_{refl} = -E_0 \cos 2\pi f_c t - (\delta E_0 \cos 2\pi f_{mod} t) \sin 2\pi f_c t. \qquad (12.22)$$

This still has the form of purely phase modulated light; the only change (compare Eq. 12.20) is in the sign of the sidebands with respect to the carrier.

Now imagine that the carrier frequency is detuned by a small amount from the resonance frequency. Depending on the context, this might be due to a shift in laser frequency, a motion of one of the cavity mirrors, or a modulation of the space-time curvature in the cavity by a gravitational wave. The light at the carrier frequency acquires a phase shift different from a pure sign flip; call it $-E_0 \cos (2\pi f_c t + \Phi_{sig})$. But the sidebands are so far from the resonance that there is no effect on their phase. The total electric field is thus

$$E = -E_0 \cos (2\pi f_c t + \Phi_{sig}) - (\delta E_0 \cos 2\pi f_{mod} t) \sin 2\pi f_c t. \qquad (12.23)$$

This is not purely phase modulated light. A graphical comparison of the difference between the reflected light described by Eq. 12.20 and by Eq. 12.23 is shown in Figure 12.7.

We can also rewrite Eq. 12.23 as

$$E = -E_0 \cos \Phi_{sig} \cos 2\pi f_c t - E_0(\delta \cos 2\pi f_{mod} t - \sin \Phi_{sig}) \sin 2\pi f_c t. \qquad (12.24)$$

Examine the second term of Eq. 12.24; the superposition of the sidebands with the phase-shifted carrier has produced a quadrature term at the carrier frequency whose amplitude depends on the difference between the applied phase dither and the phase shift in the light from the cavity. When we calculate the power of the reflected light, by squaring this expression for the reflected electric field and averaging over many cycles of the carrier, we find a component proportional to $\delta \sin \Phi_{sig} \cos 2\pi f_{mod} t$. In other words, the returning light is amplitude modulated at the frequency f_{mod}, with an amplitude proportional to $\sin \Phi_{sig} \approx \Phi_{sig}$. Thus, we have a way of producing the same result as was achieved in a non-resonant interferometer, a beam of light whose amplitude is proportional to the phase imparted to the light by the gravitational wave.

What we do next depends on the specific application we are considering. If we are locking an interferometer arm to stable laser light, then we can apply feedback through a combination of audio frequency signals applied to the Pockels cell and to a forcer attached to one of the cavity mirrors. If we are instead stabilizing a laser by locking its wavelength to the length of a good reference cavity, then our actuators need to change the light's wavelength. This can be done by physical adjustment of the laser cavity's length, such as by moving one of its mirrors. (One of the laser

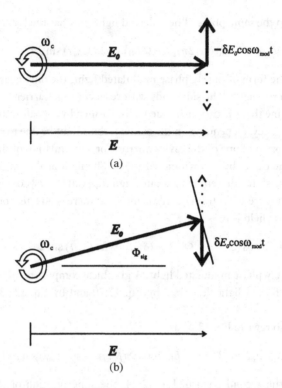

Figure 12.7:   A phasor diagram illustrating the addition of the carrier field and the modulation field: (a) with the light's wavelength centered on the cavity resonance, and (b) with the wavelength offset from resonance.

mirrors is often mounted on a small *piezoelectric translator* for just this purpose.) Rapid small fluctuations can be corrected with signals applied to a Pockels cell, as before.

12.6   A Simple Interferometer with Fabry-Perot Arms

We've remarked above that an interferometer with Fabry-Perot cavities for arms is more complex to control than a non-resonant interferometer, since control systems are necessary simply to get a useful phase response out of a Fabry-Perot cavity. But there is a clever way to construct a working gravitational wave detector that takes advantage of that fact, using only the components we've just described.[162] Construct each of the arms of the interferometer out of Fabry–Perot cavities, and equip each arm with the phase interrogation system described above. (See Figure 12.8.) Choose

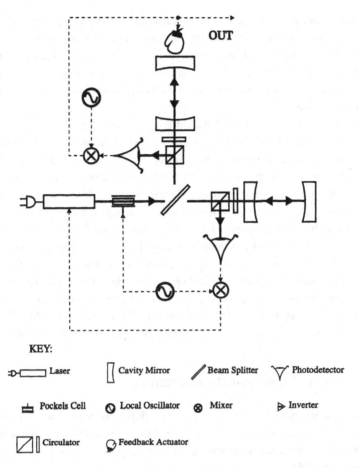

KEY:

⊐▭ Laser ⫿ Cavity Mirror ╱ Beam Splitter ⋎ Photodetector

⊥ Pockels Cell Ⓝ Local Oscillator ⊗ Mixer ▷ Inverter

◰ Circulator Ⓠ Feedback Actuator

Figure 12.8: A schematic diagram of the "simple" interferometer made with two Fabry-Perot cavities.

one of the arms to be a length reference, and lock the laser to it. In other words, use a control system to adjust the laser wavelength so that it is in resonance with the arm.

If there is a gravitational wave present, it will cause a momentary change in the effective length of the arm. With the laser locked to the first arm, the wavelength of the light leaving the laser has been transformed into a length standard as precise as the noise in the arm allows. When a gravitational wave is present, that means that the laser's wavelength is changed to match the perturbed effective length of the first arm. This matched light is also entering the second arm. Now, close a second loop, causing the second arm's length to match the wavelength of the stabilized laser. If a gravitational wave were present, it would cause an effective length shift in the second arm equal and opposite to that in the first arm. The total mismatch between the stabilized light's wavelength and the length of the second arm corresponds to twice the one-arm length change. The error signal of the second arm loop then registers the gravitational wave in the same way as the locked non-resonant interferometer of Section 12.1 above.

Both the similarities to and the differences from the non-resonant interferometer are remarkable. The scheme takes essential advantage of the fact that a Fabry-Perot cavity is a length-sensitive interferometer all by itself. Given that fact, one might wonder whether one actually needs two long orthogonal arms to make a sensitive interferometer. Unfortunately, the answer is yes. If we had tried to stabilize the laser to a cavity with a substantially shorter optical path length, then the shot noise would have been translated effectively into larger frequency noise, causing more noise in the second arm loop. If we tried to squeeze the same optical path length into a physically shorter length L, then displacement noise of the mirrors would have dominated. And if we had oriented the first arm parallel to the second (to fit them into a single long tunnel, say), then the gravitational wave signals would cancel in the two arms, instead of combining. Unfortunately, there doesn't seem to be a way to escape the basic requirements for gravitational wave detection.

12.7   Beyond the Basic Interferometer

There are a number of variants of the conventional interferometer configurations we have been discussing, whose use may lead to substantially improved sensitivity. In this section, we discuss three such techniques: *power recycling, signal recycling*, and *resonant sideband extraction*.

12.7.1   *Power recycling*

Power recycling is a technique to remedy what could be seen as a sort of profligacy in the design of interferometers of the style we've been discussing. Recall that the phase shift that a gravitational wave impresses on the light in an interferometer arm reaches a maximum when the storage time for light in the arm is half the gravitational wave period. In a long interferometer, this is achieved with a small number of reflections, perhaps as few as 30 for an interferometer optimized for 1 kHz waves. For mirrors with the high reflectivities achievable today ($1 - R < 10^{-4}$), that means that the light exiting the bright port of the interferometer is almost as bright as the light entering. What a waste! A lot of effort will have gone into constructing and taming

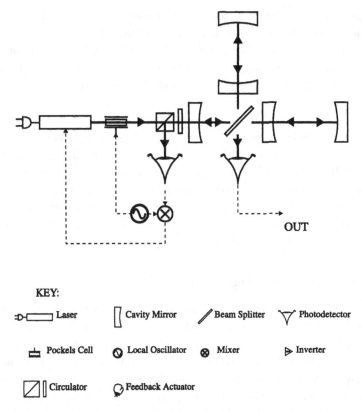

KEY:

⊃□▭ Laser   ▯ Cavity Mirror   ╱ Beam Splitter   ⋎ Photodetector

⊥ Pockels Cell   ⊘ Local Oscillator   ⊗ Mixer   ▷ Inverter

▱▯ Circulator   ↻ Feedback Actuator

Figure 12.9: A schematic diagram of a Michelson interferometer in which the light is recycled.

the high power lasers required for low shot noise. It is almost as if the bright output port of the interferometer were a second good laser, going to waste.

The idea of reusing this output light by shining it back into the input port was proposed independently in 1981 by Drever and by Schilling.[163] The analogy with a Fabry-Perot cavity becomes clear (see Figure 12.9) when one recognizes that the way to return power most effectively is to reflect it from a partially transmitting mirror in such a way that it adds coherently in phase with the fresh light from the laser. A feedback loop like the ones we've just been discussing keeps the light locked to a composite Fabry-Perot cavity: the partially transmitting recycling mirror is the input mirror, while the rest of the interferometer functions as a far mirror.

How much benefit can be achieved by this? Perhaps quite a bit. The limit to the number of "recycles" of the light is ultimately how much light is lost per trip through the interferometer. We need to keep in mind that light scattered into higher Hermite-Gaussian modes is just as lost to the interferometer as light absorbed by mirror coatings. Even accounting for this, a factor of 30 recycles seems possible, even conservative, given detailed specs for a large interferometer.[6] This would reduce the shot noise by a factor of $\sqrt{30}$, for a given available input power.

Plans along these lines point up one inadequacy of the "simple" Fabry-Perot interferometer discussed in the previous section. In that design, light from the two arms is never recombined at the beam splitter, so no light emerges from any bright output port to be available for recycling. Instead, the light is dumped onto the photodiodes used to lock and read out the arms one by one. In order to take advantage of recycling, it would be necessary to forego the simplicity of that design, and use instead one more closely analogous to the non-resonant interferometer. This is of course possible to do, at the expense of more complex sensing to keep all of the individual cavities properly locked.[164]

12.7.2　Signal recycling

There are variations on the standard interferometer design that can give particular benefits in searching for unknown periodic sources. A scheme called *signal recycling*, proposed by the late Brian Meers, uses a partly-reflecting mirror at the output port of the interferometer.[165] (See Figure 12.10.) It functions as an analog to the partially reflecting mirror of the standard recycling scheme described above. But instead of making the interferometer into a cavity resonant at the laser frequency f_ℓ, this signal recycling mirror sets up a cavity resonant at $f_\ell \pm f_{sig}$, where f_{sig} is the signal frequency. Since the gravitational wave is encoded on just such a sideband (through phase modulation of the carrier f_ℓ at frequency f_{sig}), returning this sideband allows it to be enhanced by a factor of the resonance Q.

One drawback of this scheme is that it achieves its improvements by narrowing the useful bandwidth of the interferometer. But even so, it may be our best bet for optimum sensitivity in a search for periodic gravitational waves. Note that the improvement comes for all possible source directions, unlike the Doppler demodulation schemes discussed in Chapter 14. The most practical all sky search for unknown periodic sources may consist of a set of such "resonantly recycled" measurements, each integrating a few days before the recycling center frequency is retuned.

12.7.3　Resonant sideband extraction

There is another way to implement a system whose block diagram would look like the configuration of Figure 12.10, but which is designed to enhance the broad band

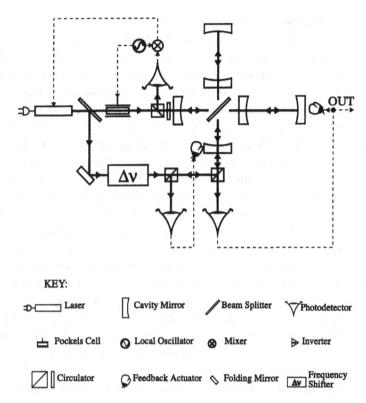

KEY:

:D— Laser	[| Cavity Mirror	/ Beam Splitter	∇ Photodetector
Pockels Cell	⊘ Local Oscillator	⊗ Mixer	▷ Inverter
Circulator	⟳ Feedback Actuator	⬦ Folding Mirror	[Δν] Frequency Shifter

Figure 12.10: A schematic diagram of a Michelson interferometer that uses signal recycling in addition to light recycling. This configuration is sometimes called "dual recycling".

sensitivity of an interferometer without requiring either high laser input power or large amounts of recycling. This is a technique called *resonant sideband extraction*, due to Mizuno *et al.*[166] It is aimed at evading the *storage time limit* in a Fabry–Perot cavity, defined in Eq. 6.31. (There is also a storage time limit in non-resonant arms, but this scheme wouldn't help in that case.)

Recall that when we use Fabry-Perot cavities in which $\tau_{stor} \gg 1/f_{gw}$, then its transfer function is

$$\Delta\phi \approx h \frac{2c}{\lambda f_{gw}}. \tag{12.25}$$

Using the language of sidebands, we can gain a qualitative understanding of this limit in a way that will suggest a clever way to evade it. A cavity with a long storage time traps the light for many round trips when it is on resonance; the total optical power travelling each way within the cavity is of order $\mathcal{F}P_{in}$, with $\mathcal{F}$ the finesse of the cavity and P_{in} the optical power shining on the cavity input mirror. A gravitational wave modulates the phase of this *circulating power*, producing sidebands whose offset from the carrier (the unmodulated laser light) is equal to the gravitational wave frequency. The strength of those sidebands is proportional to the circulating power, so in a high finesse cavity a gravitational wave creates much stronger sidebands than in a low finesse cavity. Then why do we have a storage time limit? Because we ordinarily create a high finesse cavity with an input mirror whose reflectivity is very high, so only a small fraction of the circulating light leaks out. In a cavity of the ordinary sort, the two effects precisely cancel for $f_{gw}\tau_{stor} \gg 1$, yielding the storage time limit.

The clever idea of Mizuno *et al.* is to have found a way to create, in effect, an input mirror for Fabry-Perot cavities that can have high reflectivity at the carrier frequency, but a much lower reflectivity for the optical sidebands created by a gravitational wave. This can be done by placing a partially reflecting mirror at the output port, in roughly the same location as a signal recycling mirror would go. But by choosing a different distance from the input mirrors than in the signal recycling case, one produces a rather different effect. Here, what one does is form a pair of coupled inboard mirrors for the arm cavities, each one consisting of the nominal input mirror of one of the arms combined with the sideband extraction mirror. This coupled mirror can be designed to have precisely the strongly frequency-dependent reflectivity that will let the sidebands leak out from the arms, while continuing to trap the carrier light.

Why is this a good idea? One possible limit to interferometer performance might turn out to come from limitations on the optical power passing through the glass substrates of the cavity input mirrors and the beam splitter.[142] Even small absorption coefficients would lead to unacceptable amounts of thermal expansion and thermal "lensing" (through the temperature dependence of the index of refraction) at high enough power levels. The resonant sideband extraction technique could achieve the shot noise level corresponding to the high circulating power in its ultra-high finesse cavities, without such high power levels having to pass through any mirror substrate. So long as losses at the cavity mirror surfaces are low, then thermal distortions can be minimized. This last condition does indeed appear possible with modern coating techniques.

Resonant Mass Gravitational Wave Detectors 13

Most of the rest of this book is devoted to showing how one might use interferometers to detect the effect of gravitational waves on nearly free masses. In this chapter, we turn our attention to a rather different style of detector, the *resonant mass detector*, also known colloquially as a "bar". This is the style of detector first developed by Joseph Weber, and improved substantially in the intervening years. In fact, the best detectors of this type are the most sensitive gravitational wave detectors of any variety yet constructed. This being so, you may wonder, "Why might one prefer interferometric detectors to bars?" We will defer a discussion of this question until the end of the chapter.

A nice explanation of bar detectors is given in the review article by Tyson and Giffard.[5] A clear summary of the more recent state-of-the-art is given in the article by Michelson, Price, and Taber.[167] The chapters in Blair's *The Detection of Gravitational Waves*[10] that are concerned with resonant mass detectors are a rich source of additional information.

13.1   Does Form Follow Function?

In Washington, D. C., one of the most popular places to visit is the National Air and Space Museum of the Smithsonian Institution. It is filled with famous aircraft and space vehicles. (The latter more often than not are represented by spares that never flew.) The attentive visitor can hardly avoid coming away thoroughly impressed with the courage of those who first flew these machines, and with the cleverness of those who designed them.

Almost lost among the more famous flying machines is a rather unglamorous memento of the era of "detente" during the late unlamented Cold War. This is the display of the Apollo–Soyuz Test Project (or ASTP), the point of which was an orbital rendezvous and link-up between an American and a Soviet spacecraft. On July 17, 1975, in front of television cameras, weightless astronauts and cosmonauts shook hands, exchanged gifts, and shared reconstituted food.[168]

Study this exhibit a while, and you will be rewarded with a valuable lesson. It begins with noticing how awkward and asymmetrical the ASTP looks compared to almost any of the other craft on display. Is it due to some special clumsiness of the docking adapter that allowed the vehicles to be linked together, in spite of the dissimilarity of their hatches? Perhaps then it will strike you how different the Apollo Command/Service Module is from the Soyuz capsule. The Apollo is a cone atop a cylinder, the Soyuz a sphere hooked to a cylinder with a rounded end. Apollo is polished metal, Soyuz is covered in greenish paint. Soyuz sprouts large solar panels and several spiky radio antennas, while the Apollo seems smooth and self-contained. In fact, there is hardly a single similar detail between them.

Yet, Apollo and Soyuz were designed to perform very similar functions in exactly the same environment. Isn't it true that "Form ever follows function?"[169] If it did, no one bothered to tell the American and Soviet spacecraft designers.

Of course, form has something to do with function, but it doesn't follow uniquely. The reason for pondering this example is to remind the reader that a book such as this may tend to give an exaggerated sense of inevitability to what are in fact a set of design choices. It is worth remembering this lesson as we compare and contrast resonant mass and interferometric detectors of gravitational waves. (Who knows, there may be another scheme cleverer than either of them.)

13.2   The Idea of Resonant Mass Detectors

What if the most sensitive way to measure a weak mechanical influence on a mass was by means of a sensor that needed to be in mechanical contact with the mass? (A very sensitive accelerometer, perhaps.) If we were looking for an effect due to a gravitational wave, we would need to apply such a sensor to something more extended than a point-like mass. Gravitational waves don't cause measurable effects on a single mass in isolation; that is, they don't cause a physically discernible acceleration. Rather, the effect of the transitory metric perturbation is felt as a tidal (i.e. stretching or compressing) relative force between a pair of masses or across an extended object. This relative force is related to the metric perturbation h by (recall Eq. 2.28)

$$F_{gw} = \frac{1}{2} mL \frac{\partial^2 h_{11}}{\partial t^2}.$$ (13.1)

So, we apply our sensitive accelerometer to the end of one of two identical masses that have been attached together by a spring. (In the real world, we attach the sensor to a long cylindrical mass, hence the name "bar". The gravest longitudinal mode of the bar acts like two masses attached by a spring.) A passing gravitational wave impulse would set the pair of masses vibrating about their common center of mass. That vibration, persisting long after the brief gravitational wave has passed, would register as an oscillatory acceleration in our sensor.

Is this a good idea? Yes, in many ways it is. Even the best sensor has some noise, most likely with a spectrum $n(f)$ like the generic one shown in Figure 10.3, dominated except at low frequencies by a white noise component. Imagine that the gravitational waves we are looking for are brief impulses, say from stellar core collapse. Such a signal's Fourier transform is very broad in frequency; its matched filter will thus admit sensor noise from a very wide band.

But when an impulse excites a high Q oscillator, the gravitational signal is transformed into a mechanical signal whose Fourier transform has only a narrow band frequency content. The matched filter, just the time reverse of the impulse response, admits only a narrow section of the noise spectrum of width f_0/Q. If the mechanical system resonates at f_0 with a quality factor Q, then the rms noise is only $n(f_0)\sqrt{f_0/Q}$. We could equally well use time-domain language, and say that the fact that the bar "rings" for a long time lets us average away much of the sensor noise. Thus, the high Q mechanical system makes a brief impulse more easily measurable in the presence of broad-band sensor noise.

13.3 A Bar's Impulse Response and Transfer Function

To understand the response of a resonant mass detector, we consider it with the tools of linear system theory. We can use a simple and familiar model: since the center of mass of the system is not affected by a gravitational wave, take it to be fixed. Then, all of the dynamics are contained in a simplified system consisting of one of the masses (or halves of the cylindrical bar), attached to half of the spring. (See Figure 13.1.) The spring constant makes the resonant frequency of this model system the same as that of the full bar or 2-mass system. The other end of the spring is fixed to a rigid support. This is none other than the simple harmonic oscillator that we have been considering in many guises throughout this book.

When, back in Chapter 4, we considered the mass-spring oscillator as a vibration isolator, we asked for the response of the system to a motion of the end of the spring far from the mass. As a gravitational wave detector, the relevant impulse response $g_{bar}(\tau)$ is the response to an impulsive force applied to the mass. The equation of

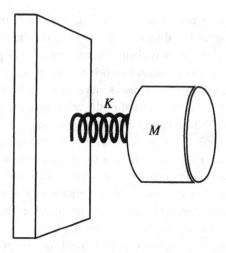

Figure 13.1: A schematic diagram of a model of a resonant mass gravitational wave detector.

motion, Eq. 4.11, yields

$$g_{bar}(\tau) = (KM - B^2/4)^{-1/2} e^{-\tau/\tau_d} \sin \Omega_0 \tau, \qquad (13.2)$$

where $\tau_d \equiv 2M/B$ and $\Omega_0 \equiv \sqrt{(K/M) - (B^2/4M^2)}$. A version of this impulse response is graphed in Figure 13.2.

What does this tell us about the response to signals other than ideal force impulses? Consider a set of "half-cycle" bursts $h(t)$, with a range of characteristic durations τ_b. The tidal force of the bar, proportional to the second time derivative $h(t)$, will then have in general a form like that shown in Figure 13.3, with balanced positive and negative sections. Recall Eq. 4.14, saying that the response of a system to a general input $s(t)$ is the convolution of that input with the system's impulse response:

$$v(t) \propto \int_{-\infty}^{t} s(\tau)g(t - \tau)d\tau. \qquad (13.3)$$

The heuristic interpretation of this rule is that each brief section of $s(t)$ can be thought of as "launching" its own impulse response. The output of the system at a time t is the superposition of all of the impulse responses produced by the inputs at all previous times.

When a burst is very short, the balanced positive and negative sections, alternating rapidly, make the net effect on the resonator small. If the pulse period is "just right", $\tau_b \sim 1/f_0$, then it will effectively excite the bar. That is because the oscillatory response of the system to each part of the brief burst is nearly in phase

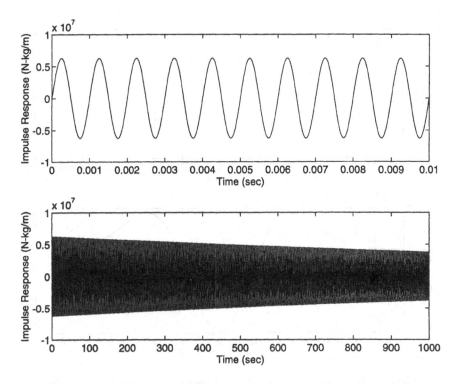

Figure 13.2:   Two views of the impulse response of a bar to an external force.

with those caused by all of the other parts. But when the burst is long, $\tau_b \gg 1/f_0$, the system response is substantially attenuated. That comes about since parts of the burst are separated by many radians of oscillation of the system. The resulting responses are out of phase, and so tend to cancel out in the superposition. Thus, a bar is most suited to detection of bursts $h(t)$ with periods comparable to its oscillation period.

The same information is also coded in the transfer function of the system

$$G(f) = \frac{1}{M(\Omega_0^2 - 1/\tau_d^2 - (2\pi f)^2) + i2\pi f B}. \tag{13.4}$$

(See Figure 13.4.) There is not much additional insight into the burst response that one can obtain from this formulation. But if we are instead interested in response to sinusoidal gravitational waves, the interpretation of the transfer function is immediate. After all, the transfer function is the complex amplitude of a system's sinusoidal response to a steady sinusoidal input of unit amplitude. As Figure 13.4 shows, a high Q resonant system has its strongest response to input signals with frequencies very

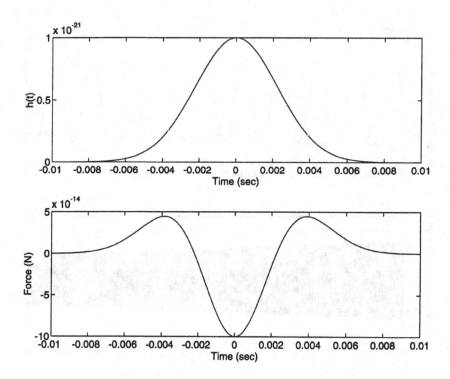

Figure 13.3: A Gaussian impulse in h, and the corresponding force across a bar.

close to the resonant frequency f_0. So, any signals with substantial fractions of their power concentrated near the bar's resonant frequency are well-suited to detection by this kind of instrument.

We'll return to the question of the signal-to-noise ratio later, after some more discussion of the noise in a realistic system.

13.4 Resonant Transducers

Weber's original sensor wasn't an accelerometer on the end of his bar, but was instead a set of piezoelectric strain gauges attached to the middle of the bar, where the strain is largest. The nearly universal switch to the accelerometer-style sensors called *resonant transducers* came about after Ho Jung Paik, then in William Fairbank's group at Stanford, invented a clever way to tailor the response of such a sensor.[170]

Recall that an accelerometer is itself a mass on a spring, equipped with a means for measuring the displacement of that *proof mass* from its rest position within its

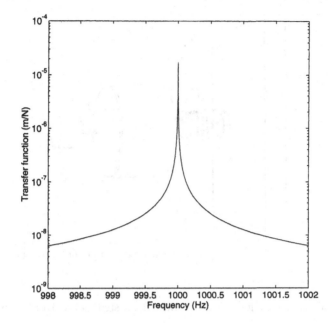

Figure 13.4: The transfer function of a bar to an external force, shown in the vicinity of the resonance.

housing. Paik's insight was that the best scheme for this application was for the transducer's resonant frequency $f_t = (1/2\pi)\sqrt{k/m}$ to be the same as the resonant frequency $f_0 = (1/2\pi)\sqrt{K/M}$ of the large bar itself. We'll analyze the response by explicitly finding the two *normal modes* of the coupled system. (For a review of modal analysis, see a text like Meirovitch's *Elements of Vibration Analysis.*[171])

We can model the bar plus transducer as a two-mass two-spring system, as shown in Figure 13.5. The matching condition between bar and sensor can be expressed as $k = \alpha K$, $m = \alpha M$, with $\alpha \ll 1$. The two coupled equations of motion of the system are

$$m\ddot{x} = -k(x - X), \tag{13.5}$$

and

$$M\ddot{X} = -k(X - x) - KX, \tag{13.6}$$

A system with two degrees of freedom, such as this one, will have two normal modes, or patterns of motion where each coordinate exhibits simple harmonic motion. Each mode has its own resonant frequency $\omega_{1,2}$, and its own *modal vector,*

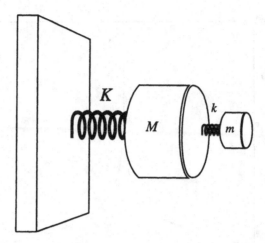

Figure 13.5:   A schematic diagram of a model of a bar with a resonant transducer attached to one end.

the ratio of the coordinates x and X when the system is oscillating at one of its resonant frequencies.

Modes are a useful concept because the solution to a set of coupled equations of motion always looks like simple harmonic motion when expressed in modal coordinates. Motion in a given mode is independent of motion in another mode. The total motion of the system is given by the superposition of its modal motions. This simplifies a problem enormously.

We start the search for a system's normal modes by first making the standard *anzatz* (trial solution) of simple harmonic motion, $x \rightarrow xe^{i\omega t}$. We plug this into Eqs. 13.5 and 13.6, and also rewrite them as a single matrix equation, to find

$$\begin{pmatrix} k - m\omega^2 & -k \\ -k & k + K - M\omega^2 \end{pmatrix} \begin{pmatrix} x \\ X \end{pmatrix} = 0. \tag{13.7}$$

The two equations of motion only have a simultaneous solution when the determinant of the matrix on the left hand side of this equation vanishes, that is when

$$(k - m\omega^2)(k + K - M\omega^2) - k^2 = 0. \tag{13.8}$$

This so-called *characteristic equation* is, for the two degree-of-freedom case, just a quadratic equation in ω^2. Thus, finding its two solutions is trivial. Those solutions give the resonant frequencies of the two normal modes.

Here, we are interested in the solutions in the limit $\alpha \ll 1$. The expressions for the modal frequencies simplify to

$$\omega_{1,2}^2 = \frac{K}{M}\left(1 \pm \frac{\sqrt{\alpha}}{2}\right). \tag{13.9}$$

Plugging these solutions, one at a time, into either row of Eq. 13.7 allows us to find the modal vectors. The two solutions are

$$\frac{x}{X} = \mp\frac{2}{\sqrt{\alpha}}. \tag{13.10}$$

Now we can see mathematically why matching $\sqrt{k/m}$ to $\sqrt{K/M}$ is such a useful choice. The coupled system will respond to an impulse with a superposition of two sinusoids of slightly differing frequencies. This means that we will observe *beats* between the modes. At its maximum, the motion of the small mass will have an amplitude $2/\sqrt{\alpha}$ times larger than the motion imparted to the "test mass" parts of the bar by the action of the gravitational wave.

If the mass ratio α is very small, the enhancement factor can be quite large. Sensor noise will usually be fixed in units of proof mass motion. If we use this trick to arrange to make the proof mass motion large, then a given amplitude of gravitational wave will be easier to see against the background of sensor noise. This trick is too good not to use. (The limit to the smallness of the proof mass comes from the *back action* of the sensor noise. See Section 13.8.)

13.5 Thermal Noise in a Bar

If sensor noise were the only noise with which we needed to be concerned, then the experimental design problem would be solved by the considerations discussed above. The matched filter is simply the time reverse of the impulse response, since the sensor noise is approximately white.

But we are, after all, considering a very sensitive mechanical experiment, so other noise sources must be considered. As with an interferometric detector, one of the most important is thermal noise. Here, though, thermal noise enters in a rather different regime. Recall that interferometric detectors are designed to have all of their mechanical resonances outside of the signal band. This lets us be concerned just with the off-resonance spectral density of thermal noise; this can be lowered by reducing the amount of mechanical dissipation in the system.

The detection strategy is fundamentally different in a resonant mass detector. The resonance is a crucial feature of the scheme for making a gravitational wave signal stand out above sensor noise. We look at the degree to which the resonance is excited, in hopes of seeing an additional excitation when a wave arrives. So anything else that can cause the resonance to be excited will be a source of noise.

Thermal noise does excite the resonance. In fact, the amplitude at which the resonance is excited is given just by the equipartition theorem. As we discussed in Eq. 7.22, the rms amplitude of oscillation due to thermal noise will be $x_{rms} = \sqrt{k_B T / 4\pi^2 M f_0^2}$, independent of the Q of the oscillator. If we just looked at the amplitude at which the resonance is excited, by applying the matched filter derived above, then this rms excitation would give a severe limit to gravitational wave sensitivity. Plugging in some characteristic numbers ($M = 10^6$ g, $f_0 = 1$ kHz, $L = 1$ m, and $T = 300$ K), we find $h_{rms} = x_{rms}/L \approx 3 \times 10^{-16}$. Using the temperature of liquid helium, $T = 4.2$ K, we have instead $h_{rms} \approx 4 \times 10^{-17}$. This isn't good enough to have a very good chance at detecting gravitational waves from astronomical sources. In fact, it would be substantially worse than achievable limits from sensor noise.

Fortunately, there is a way to substantially ameliorate this problem. We need to remember that now we are considering a situation in which the noise spectrum $n(f)$ is not white. (See Figure 13.6.) It exhibits a strong peak at the resonant frequency, due to thermal excitation of the resonance, poking up above a white spectral level due to the sensor noise. When we want to find the optimal way to extract an impulsive signal in this situation, the matched filter is no longer simply the time reverse of the system's impulse response. Rather, it is the filter whose Fourier transform $F(f)$ is given by

$$F(f) \propto \frac{G^*(f)}{n^2(f)}, \qquad (13.11)$$

where $G(f)$ is the transfer function of the bar-sensor system to a gravitational wave excitation. This is a prescription for de-emphasizing frequencies at which the noise is strong. Once thermal noise enters the picture, that means frequencies at and near the resonance.

Instead of finishing this derivation formally, it is more worthwhile to see heuristically what one needs to do. The crucial insight comes from recognizing that the vibrational state of a high Q oscillator changes only very slowly in response to a stationary noise force. It takes, on average, an interval $\tau_d \sim Q/f_0$ before the oscillator's amplitude or phase becomes uncorrelated with its initial state. It is on time scales long compared to this that one sees the full rms thermal noise scatter derived above. But on measurement time scales τ_m short compared to τ_d, the scatter is reduced by a factor $\sqrt{\tau_m/\tau_d}$.

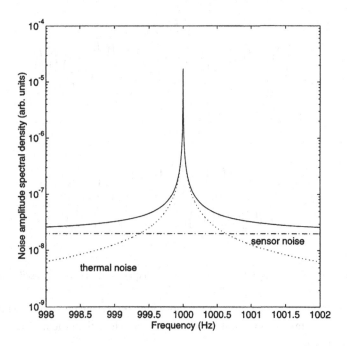

Figure 13.6: The total noise spectrum of a bar in the vicinity of the resonance. Thermal noise dominates right at the resonance, sensor noise elsewhere.

An optimization strategy should now be clear. The reason we planned on using long averaging times (convolving the sensor output with the impulse response over an interval of order τ_d) was to reduce the rms magnitude of sensor noise. If this makes the contribution from thermal noise dominate, we clearly would do better to average for a shorter time. The optimum comes at a time τ_m short enough so that the two contributions to the total noise are equal.

When the dust has settled, one finds the matched template is something like the object shown in Figure 13.7. It consists of the product of a sinusoid at the resonant frequency with a function we might call an "exponential signum",

$$D(\tau) \equiv \begin{cases} +e^{-\tau/\tau_m}, & \text{for } \tau > 0 \\ -e^{+\tau/\tau_m}, & \text{for } \tau < 0. \end{cases} \tag{13.12}$$

Applied to the sensor output, such a template looks for sudden changes in the amplitude or phase of the bar's oscillation, averaged over the optimum averaging time. It acts as a band pass filter for changes in the amplitude, rejecting the slow drift in amplitude due to thermal noise. The system, when its output is passed through a filter whose impulse response is the time reverse of this matched template, is

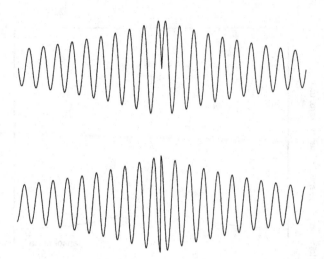

Figure 13.7:   Matched templates for the two quadratures of a bar's oscillation.

sensitive to signals with power in a band about the resonance whose width is of order $1/\tau_m$. Note that this band is typically much wider than the half-width of the bar resonance itself.

13.6   Bandwidth of Resonant Mass Detectors

For today's bars, the optimum filter prescription given above typically leads to a matched filter whose bandwidth Δf is much smaller than the resonant frequency f_0 at which the bar is most sensitive. The output of the filter is correlated over times of order $\tau_m \sim 1/\Delta f$, so it is only worthwhile to record values of its output every τ_m or so. Examination of the record of outputs is a way of asking whether, in a given interval of time of width τ_m, the bar received a significant excitation from a gravitational wave.

13.6.1   *When are narrow bandwidths optimum?*

It is worthwhile to have a heuristic understanding of why the optimum filters have, to date, always had a narrow bandwidth. One important reason has to do with the mechanical operation of the resonant transducer. Recall that a resonant transducer of small mass ratio α is chosen so that as large a motion as possible can be presented to the position sensor, to compete most successfully with the sensor's noise. When a gravitational wave burst passes through a bar, it leaves the bar and transducer system in a state in which the bar's displacement is at a maximum, but that of the transducer's small mass with respect to its sensor (fixed to the bar's end) is small. The resonant enhancement of mass motion comes about as the bar-transducer system exhibits *beats* between its two normal modes — because of the small difference in frequency between the two normal modes, their excitations shift in phase with respect to one another, until a time is reached at which the relative displacement of bar end and transducer mass are maximized. To achieve maximum sensitivity to a gravitational wave, one needs to wait for a good fraction of a beat period for the transducer mass displacement buildup to occur. Thus, the bandwidth for optimum sensitivity cannot be wider than a value of the order of the inverse of the beat period; in other words, the bandwidth of such a system will be of order the frequency splitting between the system's two normal modes.[172]

If we wanted to construct a gravitational wave sensor with a larger bandwidth at optimum sensitivity, this line of reasoning tells us that we need to increase the frequency splitting between the modes so that the beat period can be reduced. For a two-mode system, this would mean choosing a mass ratio α closer to unity. But then the resonant amplification provided by the transducer wouldn't be so great, and the noise of the position sensor would be equivalent to a larger noise in terms of gravitational wave strength. One way out of this bind has been studied in detail by J.-P. Richard.[173] In place of a transducer with a single small mass $m = \alpha M$, construct a cascaded set of N resonators, in which the ratio of successive masses is $\mu \equiv \alpha^{1/N}$; this leaves the final, smallest, mass still equal to αM. Then one would have the best of both worlds — the signal enhancement is still of order $1/\sqrt{\alpha}$, but the overall frequency splitting, and hence the bandwidth, are of order $f_0 \sqrt{\mu}$. With enough intermediate masses, the bandwidth of the system can in principle reach a good fraction of the resonant frequency. A way of optimizing designs of such multimode systems was derived by Price.[174]

Another aspect of the bandwidth problem has to do with the electrical aspects of the transducer-position sensor combination. (This combination is sometimes referred to as a *mechanical amplifier*.) We have so far referred to the position sensor as a single object whose noise was expressible in units of, say, m/$\sqrt{\text{Hz}}$. Now consider that the sensor consists of two subsystems: the first performs a transduction between mechanical motion and the modulation of some electrical property such as the inductance of a coil or capacitance of a capacitor, while the second subsystem produces an electrical signal whose value depends on that modulated electrical property, much the way an electrical *bridge* does. From this point of view, one can see that the fundamental position noise of the mechanical amplifier comes from the electrical noise of the pre-amplifier used to sense the transducer output.

The equivalent gravitational wave strength of a given amount of pre-amplifier noise is minimized if mechanical-to-electrical transduction is maximized. Clearly, one couldn't do better than to achieve a situation in which all of the bar's mechanical energy could be extracted electrically, say by dissipation in an optimally chosen resistor across the transducer output terminals, in of order one cycle of oscillation. A figure of merit, the coupling factor β, has been defined[175] as the ratio of the available transducer output energy per cycle to the bar's total energy. The transducers employed to date have usually not achieved β values of order unity; $\beta \sim 10^{-2}$ is more typical.

The relation between the coupling factor and the optimum bandwidth can now be seen: if the coupling is not optimized, then one could always reduce the significance of the pre-amplifier noise by averaging for a time substantially longer than one cycle. Thus, one is led again to a situation in which optimum signal-to-noise ratio for a burst comes at the expense of bandwidth.

Limits to increasing β stem from several factors. In some transducer designs, there exist upper limits to the electrical or magnetic field strengths that may be used, caused by breakdown thresholds in the material transducer parts. Present-day limits typically come at lower levels than this, however. It seems to be a general, although poorly understood, phenomenon for the electrical losses in the transducer to increase as the coupling factor increases. At the level at which the thermal noise from these losses begins to dominate the noise budget, there is no further advantage to be gained by increasing β.

13.6.2 *Interpreting narrow-band observations*

It is instructive to compare two situations: one in which the optimum filter has a bandwidth comparable to the signal frequency, the other in which the filter's bandwidth is small compared to the signal frequency. In the former case, appropriate when the detector noise $h(f)$ is white, the optimum filter's transfer function $G(f)$ is proportional to the Fourier transform of the signal $H^*(f)$ over a broad band. In other words, the optimum template looks like the signal waveform. (Recall Section 4.3.) By contrast, in the small bandwidth case, the optimum filter only is proportional to $H^*(f)$ over a narrow band. This is equivalent, in the time domain, to a matched template that only corresponds to a shift in the magnitude or phase of the quasi-sinusoidal output of the detection system at frequency f_0, averaged over a time $\tau_m \sim 1/\Delta f$. In this averaging process, almost all of the detail of the signal structure on short time-scales is washed out. That is to say, a large variety of different signals will give indistinguishable outputs when the optimum filter is narrow. All that matters is the magnitude and phase of the signal at the filter's center frequency.

As long as the parameters of a bar and its associated transducer and amplifier lead to a matched filter with bandwidth $\Delta f \ll f_0$, then the information that can be determined from its output is primarily the magnitude and phase of the signal's Fourier transform at the bar's resonant frequency. A suggestive way to characterize this aspect of a gravitational wave signal is to express it as the energy that such a signal would impart to the bar if it were at rest. (It can be shown[176] that this information can be obtained even from the output of a bar with significant excitation, by monitoring both quadratures of a bar's output.) An impulsive force $F(t) = P\delta(t - t_0)$ would impart a momentum P to a mass M, thus giving it an energy $E = P^2/2M$. At maximum amplitude X of the bar's oscillation, that energy appears as potential energy $4\pi^2 M f_0^2 X^2/2$, so we can also write $X = \sqrt{E/2\pi^2 M f_0^2}$. Since the displacement $X \sim hL$ (the force formula Eq. 2.28 was derived by requiring this to be so), then we can relate the energy E to the strain h by writing $E \sim 2\pi^2 M f_0^2 L^2 h^2$, or

$$h_{rms} \sim \frac{1}{\pi f_0 L} \sqrt{\frac{E}{2M}}. \tag{13.13}$$

An impulsive force makes a nice example for solving the dynamics of the bar, but it corresponds to a rather unphysical $h(t)$. Recall that $F \propto \ddot{h}$, so any function $F(t)$ that has only one sign corresponds to a strain that has the limit $h(t) \propto t$ as $t \to \infty$. Since $h \propto \ddot{I}$, this "simple" ansatz corresponds to wildly dramatic behavior of the source of the waves. So it would be nice to have a relation true for physically reasonable waveforms, even if they aren't so simple to work with. For an optimally oriented wave, the exact answer is[177]

$$E = \frac{M v_s^4}{L^2} |H(f_0)|^2, \tag{13.14}$$

where M is the total mass of the bar and L is its overall length. (The average excitation E when the wave can have any arbitrary orientation with respect to the bar is a factor $4/15$ smaller.[9]) Because the excitation state of the bar's fundamental mode is the thing that is most easily measurable about its output, a bar's response to a gravitational wave burst is most naturally expressed in terms of energy units. Often, instead of E, a temperature $T \equiv E/k_B$ is used instead. This is known as the *pulse temperature*.

Equation 13.14 gives the correct relation between the measurable quantity E and the parameters of the gravitational wave, specifically the magnitude of the waveform's Fourier transform $H(f_0)$ at the bar's resonant frequency f_0. For convenience, it is useful to be able to express the relationship in a form that refers directly to some characteristic amplitude h_c of the waveform $h(t)$ itself. We shouldn't expect an exact calibration in terms of h_{rms}; the relationship between $H(f)$ and h_{rms} depends on the details of the waveform. One standard assumption is that the waveform is a simple impulse of duration τ_g whose Fourier transform has an approximately constant amplitude over a band $\Delta f_g = 1/\tau_g$. For typical bars, we are usually interested in bursts of duration $\tau_g \approx 10^{-3}$ sec. Then a useful definition for the characteristic waveform amplitude is[177]

$$h_c \equiv \frac{1}{\tau_g} |H(f_0)| \approx 10^3 \, \text{s}^{-1} |H(f_0)|. \tag{13.15}$$

Combining this with Eq. 13.14, we have

$$h_c = \frac{\sqrt{15}}{2} \frac{L}{\tau_g v_s^2} \sqrt{\frac{E}{M}}, \tag{13.16}$$

a handy approximate calibration between a bar's natural observable E and a characteristic amplitude for the gravitational waveform.

13.7 A Real Bar

As an example of a real bar, we give here a very brief physical description of the 4 K bar that operated at Stanford University from 1981 through 1989.[167] (It is typical in many respects of others at LSU, Maryland, Rome, Perth, and elsewhere, although details differ, some substantially.)

The bar itself is a cylinder of Aluminum 5056, suspended by rods attached to points on either side of its two end faces. These rods hang from vibration isolation stacks consisting of 10 stages, designed to provide several hundred dB of isolation at the bar's fundamental longitudinal resonant frequency, near 1 kHz. The total mass is 4.8 tonnes, and the bar's overall length is just over 2 meters.

The most striking thing about the appearance of a modern bar is that it is buried within a large cryostat, arranged so that it can be cooled to the temperature of liquid helium, ~4 K. New bars at Stanford and Rome, now being tested, use dilution refrigerators to reach temperatures of order 50 mK. There are several reasons for the use of low temperatures. First is the obvious one, that thermal noise power is explicitly proportional to the absolute temperature. A less obvious reason comes from the fact that many materials, including the popular 5056 alloy, have much higher Q at low temperature, further reducing the effect of thermal noise on sensitivity. Finally, the quietest readout amplifiers for the motion sensors are superconducting devices that only function at low temperature.

Bars operating at $T = 4$ K have achieved sensitivities to 1 kHz bursts of order $h_{rms} \sim 10^{-18}$. The record so far was achieved by the LSU group, about 5×10^{-19}.[178] The new generation of 50 mK bars should go substantially below the 10^{-19} level.[167]

13.8   Quantum Mechanical Sensitivity "Limit"

Analysis of resonant bars can no more neglect quantum mechanics than can the analysis of interferometers. The uncertainly principle, $\sigma_x \sigma_p \geq \hbar$, can be applied directly to our model of a bar. We substitute for σ_p the expression $2\pi f_0 M \sigma_x$ appropriate for simple harmonic motion, and find

$$\sigma_x \geq \sqrt{\frac{\hbar}{2\pi f_0 M}}, \tag{13.17}$$

or

$$h_{rms} \geq \frac{1}{L}\sqrt{\frac{\hbar}{2\pi f_0 M}}. \tag{13.18}$$

Using our canonical parameters ($M = 10^6$ g, $f_0 = 1$ kHz, and $L = 1$ m), this yields $h_{rms} \geq 4 \times 10^{-21}$. This is somewhat discouraging. Taken at face value, it seems to say that resonant mass detectors, as currently constructed, would just barely be able to achieve an astronomically interesting level of sensitivity, if operated at the limits of performance allowed by quantum mechanics.

How does this come about? As we saw earlier, we can analyze macroscopic but sensitive measuring instruments as if they were Heisenberg microscopes, and identify, in the style of Bohr, which disturbing force is inescapably associated with a given measurement process. Recall that in the case of an interferometer, we saw that fluctuating laser radiation pressure disturbed the test masses at an unavoidable minimum level.

To analyze resonant mass detectors, we need to say a little more about the instrumentation of the motion sensor on the bar. The position of the proof mass needs to be determined by some means, ultimately converted into a large enough electrical signal to be plotted on a chart record or digitized by an ADC. Fabry-Perot interferometry is in fact one possible way to do this; J.-P. Richard at Maryland is working on this style of sensor.[179] But most bars are read out by some sort of electrical bridge circuit, in which one leg of the bridge is unbalanced by motion of the proof mass. Inductance modulation and capacitance modulation are both in use.[180]

The state of the bridge is sensed by the quietest electrical amplifier available. The best amplifiers today are devices known as SQUIDs, short for Superconducting

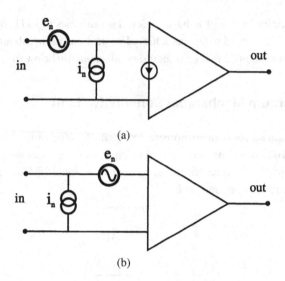

Figure 13.8:   Schematic diagrams of the noise sources in (a) a current amplifier and (b) a voltage amplifier.

QUantum Interference Devices. (The fact that a SQUID is itself a remarkable macroscopic application of quantum mechanics plays no essential role in the rest of this argument.) From an electrical point of view, the proof mass and bridge are purely passive devices; the sensor noise comes from a combination of electrical input noise in the amplifier and thermal noise (Johnson noise) in the bridge.

This noise is not just an abstract way of representing the degradation of signal as it is processed by the amplifier. There is electrical noise physically present at the input terminals of the amplifier. We can model a noisy amplifier in a way shown in Figure 13.8. For a current amplifier such as a SQUID, the model consists of an ideal noiseless current amplifier with zero input impedance, at whose inputs are two noise generators: a voltage noise source in series with the input, and a current noise source in parallel with it.[181] (For voltage amplifiers, including the ubiquitous operational amplifier, we would instead use a model consisting of an ideal noiseless voltage amplifier with infinite input impedance, as in Figure 13.8b.)

For a current amplifier, it is the voltage noise source that is the culprit. The noise that it generates acts back on the transducer, driving the electrical bridge or the electrical parts of whatever electromechanical system performs the transduction between mass motion and electrical signal. The physics of the transduction process guarantees that electrical noise injected into the transducer's output terminals will cause mechanical noise at its "input terminals", that is at the proof mass. There is a reciprocity theorem[182] that governs this interaction for a general linear transducer.

A simple example will make the connection plausible. Consider the magnetic pickup of a seismometer, in which a magnet attached to the moving proof mass causes a voltage across the open circuit terminals of a pick-up coil. This means we should read the output of the transducer with a voltage amplifier, and use the noise model of Figure 13.8b. If one were instead to inject a current into the pick-up coil, then a force would be applied to the proof mass. In fact, it is just this mechanism that is used to apply calibration signals to a seismometer, although usually an auxiliary transducer is installed for this purpose. The reciprocity theorem is expressed in this case by the fact that the *generator constant*, giving the ratio of output voltage to input velocity, is numerically equal (in consistent units) to the *motor constant*, the ratio of *back action* force on the proof mass to current injected at the output terminals.

Where does quantum mechanics come into the picture? We've now seen how the measurement process unavoidably disturbs the mechanical system being measured, but we haven't yet seen how to estimate the minimum level of disturbance. R.P. Giffard[183] supplied a proof of the required relation, extending Heffner's[151] argument about the minimum noise temperature of a linear amplifier. (Recall Eq. 11.12.) The magnitude of this noise turns out to be just that necessary to "enforce" the uncertainty principle.

Does this mean that resonant mass detectors must, like Moses, arrive at the borders of the Promised Land of gravitational wave detection, only to be barred from entering? Maybe not. The relatively high level of the quantum "limit" has made interferometry seem like a more promising method for attaining astronomically interesting sensitivity, in spite of the fact that up to the date of this writing, the best operating bar has always been more sensitive than the best existing interferometer. (At present, the margin in h_{rms} for a 1 kHz burst is of order a factor of 2.)

The key quantum mechanical difference between the two schemes is not in the details of the readout mechanism, or even directly because of the distinction between resonant and non-resonant detection. But the resonant technique is indirectly responsible. The resonance must be chosen to be at an astronomically interesting frequency. Given the range of speeds of longitudinal sound waves in materials, that in turn sets a restriction on the total length of the bar. Since quantum mechanics sets a limit to position discrimination, then a small separation between test masses (ends of bar) translates into a large quantum limit to gravitational wave strain. Interferometers, without the need to fix a resonant connection between test masses, face only the much milder restriction that there are no SNR improvements once $L_{opt} > c/2f_{gw}$. Of course as we've seen, there are less fundamental upper limits to test mass separation in an interferometer. And at any given length, there does exist a quantum limit.

In fairness to interferometers, we should briefly mention here that even if an interferometer and a bar had equal sensitivity to, say, a 1 kHz burst, it is likely that the interferometer would have an advantage in sensitivity to a wider range of signal frequencies. But then again, one could imagine constructing a "xylophone" array of bars to cover a wide range of frequencies.

Some serious thought has been given to tuning bars to lower frequencies; 100 Hz would certainly also be interesting. In addition, materials with higher speeds of sound than aluminum exist, and might be pressed into service to make longer bars.[184] Silicon carbide is one example; it has a speed of sound $v_s = 11.8$ km/sec, compared to aluminum's 5 km/sec. Such a bar tuned to 1 kHz would have a quantum "limit" of roughly $h_{rms} \approx 4 \times 10^{-22}$.

The most ambitious plan to reach sensitivity levels in the same ballpark as that of LIGO and its cousins is a proposal to build a xylophonic set of so-called truncated icosahedral gravitational wave antennas, or TIGAs.[185] The truncated icosahedron, most familiar as the shape of a soccer ball, is a nearly spherical shape whose pentagonal faces provide convenient locations to attach transducers that can sense all five of its quadrupolar modes. Compared to a bar of typical aspect ratio with the same resonant frequency, a single one of these modes has a strain sensitivity about a factor of 4 better. Another factor of roughly 2 in angle-averaged sensitivity comes from the existence of a total of 5 orthogonal modes, each oriented to be most sensitive to a different incoming wave direction and polarization. If "quantum-limited" sensors were available, then a xylophone of five such TIGA detectors might compare quite favorably to the sensitivity of an early generation broad band LIGO interferometer, in the range between 1 and 2 kHz.

13.9 Beyond the Quantum "Limit"?

From the point of view of fundamental physics, it is quite interesting to realize that it should be possible to evade the quantum "limit". (That is why we have been careful to place the word "limit" in quotation marks throughout this chapter.) The chain of thought was first pursued by Braginsky,[186] and was clearly worked out in a classic paper by Caves, Thorne, Drever, Sandberg, and Zimmerman.[187] The method, called *back action evasion*, depends on one crucial feature of an oscillator: a disturbance force applied at one moment causes a displacement response not immediately, but one quarter cycle of oscillation later.

Imagine a transducer in which we can change at will the degree of coupling between the mechanical system and the noisy electrical amplifier. An example might be an RF bridge, where we can modulate the amplitude of the RF drive signal.

If we modulate the coupling by a function proportional to $\sin 2\pi f_0 t$, then we will be able to sensitively read out the state of the oscillator in that one quadrature, but will have poor sensitivity to the $\cos 2\pi f_0 t$ term of its motion. Cleverly, the back action noise is also injected with a strength proportional to $\sin 2\pi f_0 t$, so it shows up as noisy motion only in the $\cos 2\pi f_0 t$ quadrature, when we aren't looking. There is only a small amount of noise injected into the quadrature we do observe. A demonstration of this scheme, albeit one carried out with an amplifier whose noise was substantially above the quantum mechanical minimum, was performed by Mark Bocko and Warren Johnson.[188]

This is an example of preparing the wave function of the system in a state other than the *minimum uncertainty wave packet*. This specially prepared system is said to be in a *squeezed state*. It is the mechanical analog, in fact the prototype, for the optical squeezed states that have been produced recently.[189]

At the moment, these considerations are perhaps slightly premature, since the best transducer/amplifier combinations still show noise well in excess of the quantum level. Those mechanical amplifiers will need to improve before we need to implement back action evasion schemes. In addition, thermal noise also needs reduction before the quantum limit can be reached or surpassed.

Detecting Gravitational Wave Signals 14

For most of this book, we've been considering how to make a gravitational wave detector work. Now it is time to consider what we would do with such a detector. This chapter, and the one that follows, can only offer a brief overview of the ways in which we will interpret the outputs of gravitational wave detectors. For additional information, the reader is referred to the chapter by Bernard Schutz in Blair's *The Detection of Gravitational Waves.*[197]

14.1 The Signal Detection Problem

A working interferometer is not very much like a telescope. We don't get pictures of the sky by looking through it, much as we'd like to study the universe with a new kind of radiation. The output of an interferometer is simply a single noisy time series, in which gravitational wave signals coming from nearly any part of the sky must compete for visibility with the various sources of noise we've been describing, and possibly some we haven't.

A closer analogy for a gravitational wave interferometer is a seismometer. It too gives as its output a single noisy time series. But there is an important difference as well — with good design a seismometer can easily be made so that its internal noise sources are small compared with the signals generated by motion of the ground, at least over a wide band. As you examine a seismometer output, you can be pretty certain that every wiggle you see represents real seismic motion, as opposed to detector noise. Whether a wiggle represents something interesting or not is a separate question. Most people would agree that the big wiggles caused by a large earthquake are more interesting than smaller ones.

Perhaps the closest analogy to the gravitational wave signal detection problem is that of a seismometer being used for surveillance of clandestine underground nuclear explosions.[190] There has been much argument in recent years as to whether violations of a ban on tests of nuclear weapons could be detected reliably. The characteristic seismic waves from such tests must be recognized from among a strong background of natural seismic vibrations. For seismic monitoring to be useful, there must be ways to confidently state that a particular candidate event is unlikely to be the result of a chance noise fluctuation.

For weapons test monitoring, a strategy has been worked out. A network of many seismometers spread around the globe monitors the vibrations of the Earth. Candidate events are those detected by a large fraction of the instruments within a brief interval of time, or *coincidence window*. Detailed analysis of arrival times within the window indicates the location of the source of the seismic waves. Distinguishing an underground nuclear test from an earthquake depends on analysis of the seismic waveform, in particular on the ratio of high frequency to low frequency power in the burst.

The gravitational wave detection problem is remarkably similar. We need ways to distinguish a real signal from a "bump in the night"[191] caused by uninteresting noise. The strategy that has been adopted is closely analogous to the nuclear explosion monitoring system. A global network of detectors will continually "listen" for signals from around the sky, taking advantage of the broad beam pattern of a gravitational wave interferometer. To be a candidate, a number of detectors must register a signal within a narrow coincidence window. Detailed analysis of arrival times can indicate the direction on the sky from which the signal came. Finally, detailed waveform analysis can reveal something of the nature of the source.

The strategy just described is most applicable to burst-like gravitational waveforms, but there are analogs for periodic waves and continuous gravitational noise backgrounds as well. The details of the waveform analysis are of course different in each case, and different from the seismic case as well. In the discussion below, we'll focus first on searches for bursts. This choice is justified by the fact that supernovae and binary coalescences are among the likeliest sources of detectable gravitational waves; both emit bursts of radiation. In succeeding sections, we will discuss in turn strategies for periodic sources and stochastic backgrounds as well.

14.2   Probability Distribution of Time Series

If we are to see a gravitational wave burst, it will be because the wave causes a variation in the interferometer phase difference that is large compared with the noise. So the structure of any solution to the detection problem rests on a foundation of knowledge of the probability that noise could be responsible for outputs of various sizes.

But what time series should be analyzed? It is unlikely that the "raw" unfiltered output of the interferometer is the best choice, at least if we have some idea of probable waveforms. Instead, the theorem of Section 4.3.2 instructs us to examine the time series formed by passing the raw data through a filter whose impulse response is the time reverse of the expected waveform. That is the optimal filter for the case of white instrument noise. When the noise departs from whiteness, we should instead use a filter that de-emphasizes frequencies at which the noise is high in favor of frequencies at which it is low. (See Eq. 4.29.) Since we are searching for signals whose form we won't know in detail, we will probably want to use a bank of filters to parametrize the space of possibilities.

Throughout our discussion so far, we've concentrated on describing the noise in an instrument by its power spectrum. As we saw in Chapter 4, it gives the mean noise power per unit bandwidth in the time series, as a function of frequency. Imagine that we have calibrated the output of an interferometer in terms of the strain h. If we choose a particular filter with transfer function $G(f)$ to condition the interferometer output, we can calculate the mean square noise of the filter's output by

$$v_{rms} = \sqrt{\int_0^\infty h^2(f)|G(f)|^2 df}. \tag{14.1}$$

This gives a single number that is some measure of the spread of values one would read at the filter output.

A gravitational wave $h(t)$ impinging on our interferometer would give an output from the filter of

$$v(t) \propto \int_{-\infty}^t h(\tau)g(t - \tau)d\tau. \tag{14.2}$$

Intuitively, we expect that if this output is much larger than the root-mean-square filter output due to the noise, then we can recognize it as a signal. This was the motivation for our definition of the signal-to-noise ratio in Section 4.3. The justification for our expectation that a large filter output must be due to something other than noise is that it would be highly improbable that noise alone could be responsible for an output large compared to its rms value. That is, we believe that the probability distribution for the filtered noise is negligible at large enough amplitudes. To what extent is this true?

We need to look again at how to construct useful descriptions of a noisy time series. The power spectrum (or its Fourier transform conjugate, the autocorrelation function) is a description of the degree to which a time series varies on different time scales. It does so at the expense of ignoring details of how often the time series may be found at various distances from the mean. All of that potentially rich detail has been lost by concentrating attention on the mean-square deviation. But what if we need to know more? Then we need to make a detailed study of the *probability distribution* of the time series, or of various filtered versions of it.

Imagine regularly sampling the output of the filter. Construct a histogram by counting the number occurrences N of outputs v between a finely spaced set of values, and graphing $N_i \equiv N(v_i < v \le v_{i+1})$ versus v_i. As we let the time over which we count go to infinity, and let the width of the counting intervals correspondingly shrink, the histogram becomes the total number of counted events times the probability distribution $P(v)$. That is, $P(v)$ is the probability density for the time series to have a value near v.

What form do we expect $P(v)$ to take? So far, our tools allow us to predict it should be centered on a mean value $\bar{v}$, and to have a mean-square width σ_v. Quite often, a good approximation to the form is given by a Gaussian distribution

$$P(v) = \frac{1}{\sqrt{2\pi\sigma^2}} e^{-(v-\bar{v})^2/2\sigma^2}. \tag{14.3}$$

The frequent occurrence of the Gaussian form comes about for a fundamental reason, summed up in the *Central Limit Theorem*.[192] Consider a process in which the variable of interest has a statistical character because of the influence of many small random events. The theorem states that as long as the number of those small events is large, and as long as those events are statistically independent of one another, then the probability distribution for the variable of interest will be well approximated by a Gaussian distribution. This powerful theorem is the reason so much attention is paid to the Gaussian probability distribution; it is ubiquitous.

It should not be surprising that another theorem guarantees that any linearly filtered version of an input time series with a Gaussian probability distribution also has a Gaussian probability distribution. This is convenient, since we will want to consider applying many different kinds of filters to a gravitational wave detector output.

The Central Limit Theorem holds without regard to the exact probability distribution of the individual random events causing the noise, provided that their distribution falls off "sufficiently rapidly" toward large events. If this requirement is violated, we might expect our probability distribution to depart from strict Gaussian form at large values of $v - \bar{v}$, where the Gaussian distribution would predict negligibly small probabilities. When we look for rare events, as we will in the early days of gravitational wave astronomy, it is important to remember that Gaussian statistics may give an overly optimistic estimate of how steeply the probability distribution of our noise will fall off.

For the moment, let us ignore such qualms. To the extent that a Gaussian is a good description of our filtered noise probability distribution, then we can make a definite statement about how rare are excursions far from the mean. This is such an important question to ask that a special function, the *complementary error function* erfc(z), has been defined to answer it. The probability of finding a value more than $z\sigma$ from the mean is

$$\text{erfc}(z) \equiv 1 - \sqrt{\frac{2}{\pi}} \int_0^z e^{-t^2/2} dt.$$

This function is tabulated in most handbooks of mathematical functions.[193] We show a graph of it in Figure 14.1.

Since this represents a more or less well-founded model for the rarity of large excursions from the mean, it is worthwhile recording a few values of erfc(z). The probability of a noise excursion more than $2\,\sigma$ from the mean is 0.046, more than $3\,\sigma$ only 0.0027, and more than $4\,\sigma$ only 6×10^{-5}. So our intuition about recognizing an improbably large departure from the mean is justified.

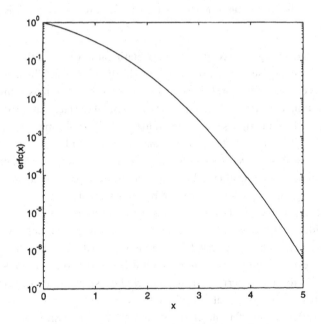

Figure 14.1: The complementary error function.

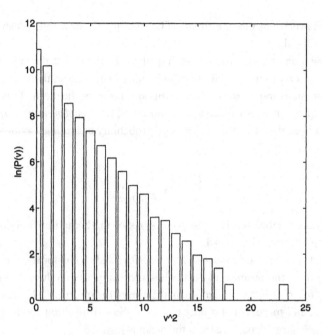

Figure 14.2: The histogram of a variable with Gaussian noise. The occurrence of a value far to the right in this diagram is highly improbable if due to noise alone; it is a candidate signal.

A sensible strategy for recognizing a gravitational wave would thus be to choose a *threshold* which noise alone is sufficiently unlikely to exceed. Then, any time the threshold is exceeded, it is probable that something other than noise caused it. Our criterion of unlikeliness must, of course, take account of the fact that we will probably be looking at the same time series through matched filters for many different kinds of signals; we'll want to be strict enough that it is really unlikely that chance alone caused the noise to exceed the threshold. If our interferometer design has been careful enough to exclude other influences, then the most probable hypothesis would be that the large output was in fact caused by the arrival of a gravitational wave. The key idea of this method is shown by the histogram of Figure 14.2.

This idea is at the heart of strategies for extracting gravitational wave signals from noise. Our defense against false claims of signal detection does, however, depend on several hypotheses about which it is legitimate to entertain doubts. One of these is that no other external influences might be responsible for large excursions. This flank needs to be continually reinforced. In Chapter 8, we examined how seismic noise could affect an interferometer, and how its effects can be minimized. A sensible precaution would be to operate several seismometers in various locations around the interferometer. One might then regard with some suspicion any large interferometer outputs that occurred simultaneously with large motions of the ground. Similarly, for any effect that might cause a spurious interferometer phase shift, the two track strategy of shielding and monitoring is advisable. (This is analogous to the '80s policy on arms control treaties, "Trust but verify.") Sensible as this strategy is, it has one weakness — it can not protect against an external effect whose influence is unsuspected. It relies on specific targeted countermeasures.

There is a second dubious hypothesis involved in the threshold-setting strategy describe above. That is the assumption that a simple Gaussian probability distribution governs the probability of the very rare events. Even if the distribution of excursions within a few σ is accurately Gaussian, how can we know the distribution continues to hold at higher levels? There are, alas, many examples of instruments in which "improbably" large excursions occur substantially in excess of the prediction from a Gaussian distribution. These may come about, of course, from the sort of external events considered above. Just as possible though, are sudden rare *glitches* within the instrument itself, especially an instrument of the complexity of a gravitational wave interferometer. Again, one might be able to make diagnostic measurements to recognize and reject putative signals caused by such effects. But also again there may be classes of such events whose causes are unsuspected, and therefore not guarded against.

Another approach entirely is possible. Why not abandon the assumption of Gaussian noise, and instead set the threshold empirically based on actual measurement of the probability distribution? This makes eminently good sense for many experiments, even though it is rather costly of observing time. The cost comes because, to avoid circularity, one needs to shield the instrument from its intended influence (run it with the "shutter closed") in order to find the rate of spurious large excursions.

Unfortunately, this strategy is not available to us as gravitational wave observers. That is because, for fundamental reasons, there is no technique available for shielding an otherwise functioning interferometer from the influence of gravitational waves. It would of course be possible to subject some subsystems, say the post detection electronics, to such tests, but it is unlikely one could gain much confidence about the whole instrument's behavior.

14.3 Coincidence Detection

What are we to do? Clearly, we need to supplement the crossing of a signal threshold with some other criterion for distinguishing a true gravitational wave signal from an impostor. The best strategy is to build more than one gravitational wave detector. Then, one is free to demand that a candidate gravitational wave signal appear in *coincidence* in several detectors. Although on the face of it this is an expensive technique, it is an extremely powerful way to guard against spurious effects of almost any type. Furthermore, we will see that it also offers the best possible remedy for a single interferometer's inability to distinguish the direction from which a gravitational wave has arrived. The great power of the coincidence method arises from two features:

1. a true gravitational wave ought to cause coincident signals in all gravitational wave detectors as long as they are all sufficiently sensitive, and
2. it is fundamentally implausible that any other effect can produce coincident signals by any mechanism other than pure chance.

One can make a confident prediction of the rate of chance coincidences based on the probability distributions of the individual instrument outputs. If the probability of a threshold crossing in instrument 1 is p_1 and is p_2 in instrument 2, then if these are due to independent causes the probability of simultaneous threshold crossings is $p_1 p_2$. On the other hand, threshold crossings due to a gravitational wave signal ought to be simultaneous, and should have nearly as high a probability as the probability to see a real event in a single detector.

The advantage gained against possible internal glitches is obtained simply by the construction of another comparable detector, no matter where it is located. The question of location becomes critical, though, when we consider rejection of unknown external influences. All else being equal, it would probably be most convenient to place the two instruments immediately adjacent to one another. Such an arrangement would be easier to manage. It would also allow the possibility of cost savings by sharing common facilities, even perhaps operating side by side in the same vacuum system. But close proximity of the instruments could easily leave them vulnerable to simultaneous disturbance from the same unknown external influence. The only remedy for this problem is to construct the instruments so far apart that correlated non-gravitational disturbances are implausible.

Wide separation of instruments is thus required, but is not without its complications. Once the light travel time becomes appreciable, it is no longer possible to expect exact temporal coincidence for a gravitational wave — depending on the direction from which the wave arrives, the time delay between real events may be up to

$$\Delta t = \pm \frac{D}{c} \approx \pm 3 \text{ msec} \frac{D}{1000 \text{ km}}, \tag{14.4}$$

where D is the straight-line distance (through the curved Earth) between the two instruments. For the two sites designated for the U.S. LIGO project in Hanford, Washington and Livingston, Louisiana, this can be up to 10 msec. Between the U.S. and Italy (location of the VIRGO interferometer), the time delay can be of order 20 msec. This slight blurring of the temporal signature of a true event must be accomodated by setting a wide enough coincidence window to accept any real event. This gives a small reduction in the statistical advantage of coincidence operation, since more random events fall in the wider window than would count as coincidences when it is reasonable to expect real events to be truly simultaneous.

A more subtle complication to using widely separated gravitational wave detectors comes from the curvature of the Earth's surface. An interferometer is only easy to build if it is installed in a nearly horizontal plane. Pendulums only offer low thermal noise for nearly horizontal arms, for example. This means that each detector points toward a different direction in the sky. To zeroth order we can ignore this, because the interferometer's response pattern is so broad. (Recall Figure 2.7.) The nulls in the response occupy only very narrow solid angles.

But at the factor of 2 level, an interferometer's beam pattern is not isotropic. This is especially true when we recall that the response is dependent on the polarization of the wave. A gravitational waveform may very well be elliptically polarized — waves coming from a binary system, for example, will have this property. This means that there are likely to be, firstly, differences in the strength of signals received by two widely separated interferometers, which may be significant if the signal-to-noise ratio is marginal. Secondly, there can be systematic differences in the waveforms received by widely separated detectors. The first is more important for detection *per se*, while the second is more important for interpretation.

14.4   Optimum Orientation

If we were going to install a second interferometer immediately adjacent to the first, then the way to match the orientation of the two is obvious. Once the two no longer share the same vertical direction, the situation is trickier. It is not hard to show that the orientation of an interferometer's beam pattern is specified by the orientation of the vector that bisects the angle between the two arms. Thus, if we wanted to match as nearly as possible the responses of two separated interferometers, then we should maximize the projection of one bisector onto the other.

The matched orientation is the optimum, if matching the reponse of the interferometers is our objective. This seems the conservative strategy to adopt if we want to maximize the chance we will see some coincidences, without additional knowledge about the properties of the signal waveforms. As pointed out by Schutz,[194] though, binary coalescence signals might call for a different strategy. Since these signals are elliptically polarized, matching the polarizations of two interferometers might be less important than the advantage one might gain in sky coverage by deliberately mismatching the orientations.

14.5   Local Coincidences

We have seen that coincident detections from widely separated instruments will be required if we are to have confidence that an apparent gravitational wave detection is not merely a terrestrial "bump in the night". But since not every real wave will cause detectable coincidences (because it may arrive from an insensitive direction for one of the detectors), we have not yet arrived at a perfect solution to the discrimination problem. The expensive but fundamentally correct strategy is to construct a network of four or more detectors. This will also pay substantial dividends both for waveform study and for source position determination, as we will see in the next chapter.

In the arsenal of discrimination techniques, we should not neglect the possible benefits of *local coincidences,* coincidences between two instruments constructed at the same location. We criticized their utility above because they might be caused by common external influences. This may be so, but they have one great advantage over remote coincidences — they can only come from a gravitational wave if the match in time and in waveform is exact, within the precision allowed by measurement noise. At a minimum, one could imagine obtaining an excellent veto signal from a local coincidence requirement; unless two co-located interferometers gave matching outputs, the cause cannot be a gravitational wave.

One interesting variant of this technique, proposed by Drever,[195] has been incorporated into the design of the LIGO project. Provision has been made for the construction of a *half-length interferometer* at one site, sharing the same vacuum system as the main interferometer. Such an interferometer could be constructed with twice the finesse as the full-length interferometer, and thus the same storage time and shot noise sensitivity. It would, though, register motions of tests masses due to displacement noise with twice the amplitude as the full-length interferometer. Internal glitches in an interferometer should only appear in one interferometer. When a gravitational wave of sufficient strength arrives at this pair of instruments, the shorter interferometer will register a signal equivalent to signal in the full-length interferometer. It would only be by a strange chance if locally generated disturbances gave the same ratio as a gravitational wave; the mean ratio should be 2 to 1.

The signal must thus satisfy the strict requirements of local coincidence, with this twist: only a phenomenon with the "tidal" character of a gravitational wave will provide signals of the required ratio in the two interferometers. It thus makes use of another characteristic of gravitational waves to help us recognize the real thing when we see it.

14.6 Searching for Periodic Gravitational Waves

When we want to search for steady periodic gravitational waves, a rather different set of considerations applies than those we've discussed for the case of impulsive waves. Questions of the rate of signal arrivals do not arise here — by assumption this class of possible signals is always present. And, rather than needing a bank of templates to match against the interferometer output, to first order a sinusoid at the fundamental frequency of the periodic wave will be a good way to search for unknown signals of this class.

This means that a good first cut at data analysis devoted to signals of this class is simply to compute the Fourier transform of the interferometer output, and to look for frequencies that show up in the spectrum at improbably strong levels. The same signals of course must appear in any interferometer that is "on the air"; this is the equivalent of the requirement of coincident detections of impulsive gravitational waves.

For the material in this section, good references are Livas's Ph. D. thesis[196] and the chapter by Schutz in Blair's book.[197]

14.6.1 *When is a spectral peak improbably strong?*

If we could assume that the noise power spectrum were white and stationary, then the probability distribution for the magnitudes v_f of individual frequency bins f in a power spectrum of the time series would be a *Rayleigh distribution*, given by[198]

$$P(v_f) = \frac{v_f}{\sigma^2} e^{-v_f^2/2\sigma^2}. \tag{14.5}$$

The statistical problem for periodic signal detection would then be to choose a threshold level so high that any bin whose magnitude exceeded the threshold is unlikely to have done so simply because of the random noise.

Experimental verification that this probability distribution holds for the spectrum of an interferometer output was provided by Niebauer *et al.* for the 30-meter interferometer at Garching. Unfortunately, there can be reasons that other instruments' spectra may not accurately obey a Rayleigh distribution, just as there can be departures of a time series from Gaussian statistics. Here, the main reason is that white power spectra from one noise source (such as shot noise) are often seen in superposition with other noise terms containing sharp spectral lines. One possible source could be, in our case, mechanical resonances in parts of the interferometer, excited by thermal noise or external vibrations. Livas' analysis of the spectrum of data from the 1.5 meter interferometer at MIT shows an example of this effect.[196] Without additional information beyond simple statistics, it would be hard to wade through the instrumental structure in the spectrum to find the signals hiding in the noise.

14.6.2 *Signatures of periodic gravitational waves*

Fortunately, there are several unambiguous features that a true periodic gravitational wave signal should show which are unlikely to be mimicked by any instrumental effect. The requirement that the same peak be seen in several instruments is quite powerful, even for two instruments constructed from the same plans. Recall that the frequency resolution in a Fourier transform of a time series of length T is $\Delta f \sim 1/T$. Thus, a transform of even a modest length of data can give excellent fractional frequency resolution. Resonances in two "identical" instruments should be close in frequency, but will not be exactly the same. On the other hand, precisely the same gravitational wave will interact with the two detectors.

There is a second way to distinguish a periodic gravitational wave from an instrumental resonance that is so distinctive it would almost certainly be convincing even if only applied to the data from a single inteferometer. The idea is to look for the amplitude and frequency modulation expected for an extraterrestrial wave impinging on detectors fixed to the moving spinning Earth. This signature also brings an additional benefit — it allows an accurate determination of the position of the source on the sky, even with the signal from only one detector.

Let's see how this might work in practice. Assume for the moment that we know we are looking for a source located at some particular location on the sky, specified by its right ascension α_s and its declination δ_s. As the Earth rotates, each location on the sky "moves" about the interferometer, appearing in a sequence of azimuth and elevation angles with a period of one sidereal day. Since the interferometer sensitivity pattern varies at about the factor of two level as a function of angle, we expect a periodic and characteristic amplitude modulation of the received signal. And, because the sensitivity pattern of an interferometer depends on the polarization of the wave, then we could determine this parameter as well from the signals received by a single instrument.

Even better position determination is available by taking into account the *frequency modulation* of the signal due to the Doppler effect. The frequency of the received wave is

$$f = f_0 \left(1 + \frac{\mathbf{v} \cdot \hat{\mathbf{r}}}{c} \right), \tag{14.6}$$

where f_0 is the "true" frequency of the signal, $\mathbf{v}$ is the velocity of the interferometer with respect to the source, and $\hat{\mathbf{r}}$ is the direction to the source. To first order, the velocity of the detector can be modelled as the resultant of sinusoidal motions due to the Earth's rotation once per sidereal day and revolution about the Sun once per year. The relation of the frequency modulation to the source position enters through the product $\mathbf{v} \cdot \hat{\mathbf{r}}$.

It is an interesting exercise to compare the scale of daily and yearly frequency modulation with the frequency resolution attainable in Fourier transforms of data sets of various lengths. The Earth's mean orbital speed v_{orb} is 2.98×10^4 m/sec, or $v_{orb}/c = 10^{-4}$. Its rotational speed v_{rot} at a latitude of 40 degrees is 355 m/sec, or $v_{rot}/c = 1.2 \times 10^{-6}$. One day contains 86,400 seconds, so if we were to compute a transform from a data set one day long, we could resolve frequency shifts of 1.2×10^{-5} Hz, or a fractional shift of about 10^{-7} in the vicinity of $f = 100$ Hz. Thus in a single day's integration, it would be easy to resolve the Doppler shift in the received signal due to the Earth's rotation. Similarly, denoting the direction shift in the Earth's orbital motion after one day as $\Delta\theta$, the shift in frequency due to the orbital motion is $\Delta\theta \times v_{orb}/c = 1.7 \times 10^{-6}$, also easily resolvable with one day's integration.

As frequency resolution improves with a lengthening data set, angular resolution of the source position on the sky grows as well. We can make an order of magnitude estimate of the angular resolution by noting that the position of the source $\hat{\mathbf{r}}$ enters the Doppler shift in the combination $\mathbf{v} \cdot \hat{\mathbf{r}}/c$. This means that the position error determined by the frequency shift must be of order

$$\frac{v}{c} \Delta\theta \sim \frac{1}{fT}, \tag{14.7}$$

or

$$\Delta\theta \sim \frac{c}{vfT}, \tag{14.8}$$

with $\Delta\theta$ in radians. (Note, on time scales of weeks or longer, the orbital modulation goes through a significant fraction of a radian, and is thus the dominant term. For time scales of order a day, the orbital term must be discounted by the fraction of a cycle through which the modulation runs. Then, taking v to be of order v_{rot} gives about the right answer.)

For integration times of 1 day, we have $\Delta\theta = 0.15$ rad, or about 10 degrees. For integration times of 2 months, when the orbital Doppler shift dominates, the coordinates of the source could be pinned down to of order 20 μrad, or about 4 arcsec.

These remarkable numbers mean that, if we can detect a periodic source of gravitational waves, then it will be possible to pin down its position with some accuracy. It might very well be possible to obtain an electromagnetic identification of the source if it has any electromagnetic counterpart, and thus be able to study it using the full panoply of astronomical techniques.

But this promise comes coupled with rather substantial computational challenges. The key difficulty comes from a natural wish to also take advantage of a long integration time to improve the signal-to-noise ratio in a search for faint sources of periodic gravitational waves. When trying to extract a purely sinusoidal signal from a white noise background, the amplitude SNR should grow as $\sqrt{T}$. This is because in a transform of a time series of length $1/T$, the frequency resolution scales as $1/T$; the noise amplitude in the bin that contains the signal is proportional to the square root of the bin's bandwidth. It is precisely this scaling of SNR with T that is disrupted by the Doppler shift; once T is long enough that the shift can be resolved in the spectrum, there will be no further improvement of SNR with T unless we make a specific correction for the Doppler shift.

If we are blessed with signals strong enough, then this will not pose too much of a problem. Instead of a single peak in a spectrum, we expect instead to see the characteristic carrier plus sidebands of a frequency modulated signal. We discussed the characteristics of the related phase modulated sinusoid in Chapter 6. Here, though, we are in a rather different regime. We consider a signal of the form

$$h(t) = h_0 \cos\left(2\pi f_0 t + \delta \cos\left(2\pi f_m t\right)\right). \tag{14.9}$$

Here f_0 is the gravitational wave frequency (playing the role of the carrier), f_m is the modulation frequency (either the daily frequency of 1.2×10^{-5} Hz or the annual frequency, 3.2×10^{-8} Hz), and δ is the dimensionless *modulation index* given by

$$\delta = \frac{f_0}{f_m} \frac{(\mathbf{v} \cdot \hat{\mathbf{r}})_{max}}{c}, \tag{14.10}$$

proportional to the maximum value of $\mathbf{v} \cdot \hat{\mathbf{r}}$ over the modulation cycle.

The character of the spectrum of sidebands depends on the value of the modulation index δ, since the amplitude of the nth sideband is $\mathcal{J}_n(\delta)$, where $\mathcal{J}_n$ is the nth Bessel function of the first kind. If $\delta \ll 1$, the spectrum would be dominated by power at the carrier frequency, with substantial power only in the first-order sidebands at $f_c \pm f_m$. This is the applicable case, by choice, for the modulation used to lock an interferometer, as discussed in Chapter 12. Here, though, we are in the opposite regime, $\delta \gg 1$. In this case, power is divided roughly equally into about δ sidebands, since in this limit the Bessel functions have an asymptotic form $\mathcal{J}_n(\delta) \sim \sqrt{2/\pi\delta}\cos(\delta - \pi/4 - n\pi/2)$.[199] Thus each peak has an amplitude of order $1/\sqrt{\delta}$ of the amplitude of the unmodulated wave.

The existence of this modulated spectrum in place of a pure sinusoid has a substantial effect on the signal-to-noise ratio in a long duration experiment. The daily modulation term gives $\delta \sim 10$ sidebands on a signal of $f_0 \approx 100$ Hz, each of amplitude roughly 1/3 the original peak. Thus a naive peak search algorithm has suffered a factor of 3 reduction in SNR.

An experiment carried out for $T \sim 1$ year suffers a much greater degradation over what one could have achieved if modulation were not present. There, $\delta \approx 3 \times 10^6$, so the spectral peaks are reduced in amplitude by a factor of $\sqrt{3 \times 10^6}$ below that of a single unmodulated sinusoid. If there were no way to redress this problem, there would appear to be virtually no benefit to carrying out a long integration to improve sensitivity. And this may be crucial, since it is unlikely we can count on signals strong enough to be seen in short integrations.

There is, in fact, a way to recover all of the information that has been scrambled by Doppler modulation. If we know the location of the source on the sky, then the factor $\vec{v} \cdot \hat{r}/c$ can be calculated exactly throughout the integration interval. This can then be used to predict the precise modulation pattern, allowing the recovery to be done with a template specialized to that modulation pattern alone.

There is a simpler way to visualize how such a scheme might be carried out. We can just as easily calculate the position of the detector in a frame "fixed to the stars" as we can its velocity. Using this trajectory, we can resample the interferometer time series in a way that corresponds to evenly spaced intervals of time for a wave travelling in a particular direction. That is, we resample the time series in units, not of t, but of

$$t' \equiv t + \frac{\Delta x \cdot \hat{r}}{c}, \tag{14.11}$$

where Δx is the interferometer's position as a function of time. Once this is done, a Fourier transform of the new time series collapses all of the sidebands into a single bin. The signal-to-noise ratio of the naive estimate is restored. Livas (1987) developed and carried out this procedure in his Ph.D. thesis.[196]

This procedure, although it takes a bit of work, is quite beautiful. Note that it works for all possible signal frequencies simultaneously; the rescaling of the time makes no reference to frequency. There is just one drawback to it — it only works for one location on the sky at a time.

Unfortunately, the precision with which one must specify the assumed position on the sky is quite fine. In fact, it is the same as the precision with which we said we could measure the position by the Doppler effect in Eq. 14.8. What was a blessing in the high SNR case is a curse in situation where one wants to use a long integration to enhance a weak signal.

Of course, if we actually know where we expect a source to be, this is not a particularly extreme burden. More likely, we might have a list of interesting positions, and use this technique to search each of them, one at a time. So far, no one has come up with computationally tractable way of carrying out this procedure for the whole sky simultaneously, although Schutz has come up with a technique that reduces the computational burden by "stepping around the sky" gradually.[197] The number of distinct cells on the sky can be calculated using Eq. 14.8. For a one year integration the number is so large that, out of fear of shocking the reader, I refrain from writing it explicitly.

14.6.3 *Frequency noise in the source and elsewhere*

An implicit assumption in the preceding discussion of the problems of long duration spectra is that the source of gravitational waves emits a rock-steady tone, constant to of order 1 part in 10^{10}. Remarkably, we expect that some objects, notably pulsars, could have such constant frequencies. In fact, even the most dramatically slowing solitary pulsars have fractional frequency changes this small, with noise about the secular slow-down smaller still.[200]

With other interesting sources, we are less lucky. Any object within a binary system will emit waves that arrive at the Solar System frequency modulated by exactly the same mechanism that causes an Earth-based detector to add the frequency modulation discussed in the previous section. Unfortunately, this includes the "Wagoner stars" that may be emitting strong periodic waves due to the Chandrasekhar-Friedman-Schutz instability. If we have a good ephemeris for the source motion, we can also correct for its motion with the same techniques just described. Otherwise, long integration times won't aid in detecting gravitational waves from such sources.

14.7 Searching for a Stochastic Background

Several mechanisms have been proposed for the generation of a background of gravitational wave noise. Most plausible are events in the early history of the Universe;

these should produce backgrounds that are accurately isotropic. It is a significant experimental challenge to convincingly demonstrate the existence of radiation with no distinguishing temporal or spatial features. These difficulties are illustrated by the story of one of the greatest achievements of modern astronomy, Penzias and Wilson's discovery of the cosmic microwave background radiation.[201]

Arno Penzias and Robert Wilson were young radio astronomers in 1965. They had been hired by Bell Laboratories to study sources of cosmic radio emission; this was interesting to the phone company as a possible background that could interfere with satellite communication systems. The standard method for observing individual bright radio sources, alternately pointing the telescope toward and away from the source, fails in the *confusion limit* where sources crowd together on the scale of the beam size. But Penzias and Wilson had equipped their radio telescope with a *cold load*, a radio black body that could be connected to the receiver in place of the antenna, and whose temperature could be accurately set to be near the expected antenna temperature. In short, they had constructed a version of a Dicke radiometer, such as we discussed in Chapter 10. Switching between cold load and antenna was performed by hand; the purpose was less to combat $1/f$ noise in the receiver than it was to permit accurate absolute calibration of the strength of the radiation of the sky, without relying on motion of the antenna pattern.

It was precisely this capability for absolute calibration that enabled Penzias and Wilson to recognize the 3 K cosmic microwave background radiation, in spite of its isotropy to 1 part in 10^3. The first time they turned their receiver to the sky it showed that the background was present. But in spite of the excellent design of their instrument, they did not believe that the signal they saw was real when they first saw it. It took a year of careful systematic checks, and the news that the background had been predicted, to give them the confidence to publicly announce their results. This story shows how difficult it can be to recognize an isotropic background of noise, even with careful preparation.

At first sight, it would seem that recognizing a gravitational background would be harder still than the microwave background. After all, we have no way of replacing the "antenna" with a gravitational black body for calibration, let alone, even more crudely, to simply block the input to see if the noise goes away. Of course, we will be able to calibrate by slightly less direct means, say by a force applied to one of the test masses. But we will miss the other advantage of a Dicke-switched instrument, the ability to distinguish external (and thus interesting) noise from noise simply added by the receiver.

But gravitational wave observations are blessed by other features that should make unambiguous identification of a stochastic background possible, as long as it is sufficiently strong to detect. These features reflect the fact that we will be looking for signals in the audio frequency band instead of the microwave band. This single fact helps in two ways. Firstly, it is what makes it possible to digitize and record the actual gravitational waveform $h(t)$. By contrast, radio receivers usually measure power, the envelope of the electric field waveform. This is often recorded only at a bandwidth low compared to the radio frequency itself.

The second advantage of the low frequency of our signals is that we can compare signals from independent and widely separated receivers, which are nevertheless separated by a distance comparable to or smaller than a wavelength. (Recall that a 10 Hz wave has a wavelength of 3×10^4 km, while a 1 kHz wave has $\lambda = 300$ km). This is both crucial and distinctive. It means that even though the interferometers may be bathed in signals from all parts of the sky, $h(t)$ will be substantially the same at each instrument. Then, so long as the interferometers are also close enough that they are oriented in substantially the same direction, the response of the two instruments should be the same. Thus, the signature of a gravitational wave background is "noise" that matches in two or more independent detectors.

Given this reason to want to correlate outputs of well-aligned interferometers exposed to the same gravitational wave field, it would seem that two interferometers at the same site would be ideal for this sort of measurement. That would be true, if only one could be certain that there were no other plausible mechanisms that could cause correlated outputs. This is the same worry we discussed for the case of impulsive signals. Here, as in that other case, it is hard to see how one could ever be confident in a claimed gravitational correlation between signals in co-located interferometers. But, as in the impulsive case, one can confidently use local correlation as a veto to check correlations between well separated interferometers. If there is a real gravitational background it must show up in local correlations.

If network designers were primarily concerned with doing the best job of detecting a stochastic background, they might insist that separate interferometers not be more than, say, 300 km apart, so that performance would be optimum up to frequencies of 1 kHz. The LIGO project's preference for a pair of sites separated by continental distances reflected the judgement that resources are best deployed to support searches for impulsive signals. If two interferometers are installed in Europe, they will likely be substantially closer together, and hence the pair best suited for stochastic background searches.

The data analysis procedure for a background search is simple; compute the zero–offset cross-correlation between two interferometer time series

$$s_1 \star s_2(\tau = 0) \equiv \lim_{T \to \infty} \frac{1}{T} \int_0^T s_1(t)s_2(t)dt. \tag{14.12}$$

But what pre-filtering should be applied to the data? There is certainly no benefit to including frequencies that are very noisy. Scaling of the signal-to-noise ratio with bandwidth is different here than it is when one is looking for a particular (deterministic) signal in the noise. Here, in effect, one noisy time series is the template for the other time series. Thus, SNR only scales as $(\Delta fT)^{-1/4}$.[197] In a time series with a non-white power spectrum, one typically does better to filter the data with a narrow filter that passes only the quietest section of the spectrum, as long as one expects that the background spectrum has no strong features. For more details on this aspect of data analysis, see the work of Christensen.[202]

Gravitational Wave Astronomy   **15**

Here we finally come to a discussion of the payoff for all of the hard work described in the preceding chapters — interpretation of detected gravitational wave signals. We start the chapter with a discussion of the determination and interpretation of source positions on the sky, because of this topic's fundamental importance in linking gravitational wave observations with the rest of astronomy. This is followed by a survey of what might be learned from the interpretation of gravitational waveforms themselves. We conclude the chapter with a capsule summary of the results of gravitational wave astronomy to date.

15.1   Gravitational Wave Source Positions

In the previous chapter it almost seemed as if we were forced against our will to propose simultaneous observation using a global network of gravitational wave detectors. In fact, such a network is precisely what one would design to overcome the single most serious difficulty with an interferometer's broad beam pattern: its inability to distinguish from which part of the sky a gravitational wave arrived. Correlation of signals will link the detectors in the network into a single gravitational wave observatory, with both nearly full sky coverage and, in principle, good position resolution as well. What was a drawback for the coincidence discriminant becomes an advantage here; differences in signal arrival times carry the information about the wave's propagation direction. Position determination becomes an exercise in triangulation on the celestial sphere.

By cross-correlating the outputs of two detectors, one can determine a best estimate of the time shift Δt_{sig} between the two signals. Real-time correlation is not necessary for this purpose, but good time-stamping of each data stream is. A single two-detector time difference defines a circle on the sky from which the wave might

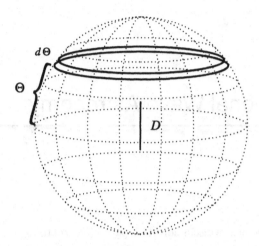

Figure 15.1: A diagram of the celestial sphere (seen from outside!), on which is drawn the position circle from the difference in the arrival times of a signal at two gravitational wave detectors.

have come. (See Figure 15.1.) This is generally not a great circle, but more like a parallel of latitude or declination in a system whose polar axis is an extension of the baseline of length D connecting the two detectors. The "declination" angle of the *position circle* is given by

$$\Theta = \arcsin\frac{c\Delta t_{sig}}{D}. \tag{15.1}$$

Observations with finite precision define a strip of width $d\Theta$ instead of a circle on the sky. Clearly, there is a great benefit that accrues from accurate time difference determinations. Achieving this will put substantial demands on the SNR to minimize random errors, and on waveform and polarization determination to minimize systematic errors.

Three detectors give two independent time differences. The two independent position circles will intersect at only two patches on the sky. (See Figure 15.2.) A fourth observation would give a third time difference, enough to distinguish a single patch on the sky. If we had a network of detectors separated by baselines of order 6×10^3 km (the radius of the Earth), and if time resolution of 0.1 msec were possible, then the position of the source would be determined to about $(5 \text{ mrad})^2 \approx (17 \text{ arcmin})^2$. That is quite respectable angular precision.

With consideration of the waveform as well, more is possible with less. Consider the case of a wave known to be linearly polarized. For example, axisymmetric sources will generate linearly polarized waveforms; stellar core collapse is a likely source.

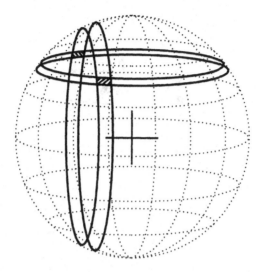

Figure 15.2: Detection of a signal by three instruments gives two time differences. The two error circles overlap in two position error boxes.

Any interferometer, regardless of its orientation, will detect the same waveform, up to a single multiplicative constant. The set of signal ratios between interferometers depends on where the source falls in the various beam patterns. Thus the ratios give extra information that can help determine the source location, even when an insufficient number of time differences have been measured.

The more general case of elliptical polarization is subtler, but the principle is the same. A very extensive discussion of simultaneous position and waveform determination is given by Gürsel and Tinto.[203]

15.1.1 *Network figure of merit*

A simple figure of merit can be calculated for the arrangement of detector sites in a network of up to four detectors. What we want is a measure of how well the set of sites is disposed for determining the angular positions of gravitational wave sources. The idea is straightforward. With two sites, there is a single distance or *baseline* between them — the larger is its length, the better is the discrimination in the angle Θ defined in Eq. 15.1 above. If we add a third site, we would clearly do best to try to construct as long a baseline as possible in a direction orthogonal to the first. In that way we construct a "position circle" in a direction orthogonal to that generated by time differences measured with the first baseline; the longer is this perpendicular baseline, the narrower is the second circular strip, and the smaller the area of the

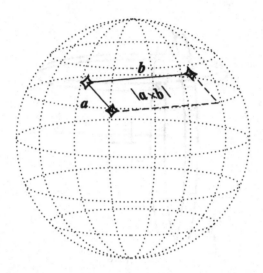

Figure 15.3: The locations of three gravitational wave detectors on the terrestrial globe, illustrating the area figure of merit for a network.

error boxes formed by their intersection. A quantity that is proportional to the length of one baseline and to the perpendicular component of the other is the area of the triangle whose vertices are the three sites. Or, if we designate one site as the origin, and write the vectors to the other two as **a** and **b** respectively, then the natural figure of merit is

$$\mathcal{A} \equiv |\mathbf{a} \times \mathbf{b}|, \tag{15.2}$$

the area of the parallelogram with sides **a** and **b**.(See Figure 15.3.)

Adding a fourth site to the network can give additional discrimination in a direction perpendicular to the plane of the first three detectors. A measure of the joint lengths of the mutually perpendicular baselines is the volume of the box defined by them, or

$$\mathcal{V} \equiv \mathbf{a} \times \mathbf{b} \cdot \mathbf{c}, \tag{15.3}$$

where the vector from the first site to the fourth is **c**. (See Figure 15.4.)

We have implicitly assumed above that sources of gravitational radiation are distributed uniformly across the sky, so that no overall orientation of baselines is preferred. The situation may be otherwise, especially if, as is possible, the Virgo Cluster may harbor the bulk of detectable gravitational wave sources. If supernovae are a strong source of burst events, the Virgo Cluster is where we'll find most of the strongest ones, as we saw in Chapter 2. On the other hand, if coalescing neutron

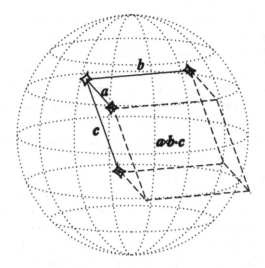

Figure 15.4: The volume figure of merit for a network of four gravitational wave detectors.

star binaries turn out to be our best targets, we'll look substantially deeper, and thus to a more isotropic distribution of galaxies as well. Perhaps it is fortunate that the latter set constitutes our single most important target, since the actual geographical distribution of the worldwide network of interferometers is as likely to be determined by the countries in which the scientists (and their sponsors) live as by a desire to "aim" a network at Virgo.

15.1.2 *Why measure positions?*

We want to determine the propagation direction of any gravitational waves we detect for a number of reasons. Firstly, it is a natural by-product of checking for coincidences between detectors. Once we've verified that all the threshold crossings fall within the expected coincidence window and thus constitute a candidate gravitational wave signal, position determination comes almost for free.

Another level of consistency check is possible from a position determined by timing data. Source position and polarization together determine the signal amplitude ratios expected from the various detectors. Thus, position determination is part of the process (perhaps iterative) by which we try to determine what gravitational waveform might have been responsible for the set of detected signals. If no waveform is consistent with all of the detector outputs, then we cast doubt on the hypothesis that a gravitational wave has been detected.

There is also a qualitatively new kind of information that will be revealed by position measurements. Assume that by various means we have been able to convince ourselves that we have detected a genuine gravitational wave. The very next question we would want to ask is the astrophysical question: "What sort of object out there generated this wave?" If the recent history of astronomy is any guide, then finding some link between a detected gravitational wave and objects visible by more conventional astronomical methods may provide crucial insights. To forge this link, few clues could be more powerful than to match the position of the gravitational wave source with that of a distinctive object about which much is known or can be learned by other means.

There are several levels at which the correspondence between gravitational and other observations might be achieved; these give payoffs of different sizes. At its most fundamental level, the exploration and explanation of a new astronomical phenomenon is confronted at its outset with a crucial challenge: to determine the true physical scale of the object in question. By determining the distance to an object, we can tell if the signal we receive is coming from a weak source of radiation that is nearby, or whether instead it comes from a very luminous object far away. Interestingly, a crucial clue for estimating an object's radial coordinate is its position in the sky, especially if that position places it in apparent proximity to other objects of known distance.

Even crude angular coordinates might be enough to indicate the mean distance of a set of sources. The discovery of galactic X-ray sources was an episode showing the power of the method. Even before precise positions showed them to be compact objects in binary systems with visible stars, a map of the locations of the sources on the sky showed an unmistakable match with the band of the Milky Way. (This kind of match would be possible with position errors of several degrees or more.) This fact demonstrated convincingly that the typical source was at a distance of the order of a few kiloparsecs, and let the true luminosity be estimated to an order of magnitude.[204]

It is certainly possible that the same pattern might emerge in the first hundred gravitational wave source positions to be determined. It is considered somewhat more likely, though, that a map of the strongest events would look like a map of the Virgo Cluster, which spans a region 10 or more degrees in diameter, centered at right ascension $= 12^h 30^m$, declination $= +13°$.[205] If this pattern should emerge, it would give an immediate indication that the typical gravitational wave source in this set was at a distance of 10 to 20 Mpc.

At almost any distance scale from 100 pc to 100 Mpc, a characteristic pattern ought to emerge from a map of gravitational wave source positions; such a pattern would indicate with what visible inhomogeneities of the cosmic distribution of matter the gravitational wave sources are associated. A difficult puzzle might be presented, though, if the distribution appeared to be isotropic. This is precisely the dilemma that has confronted the study of gamma ray bursts since they were discovered in 1973.[206] These enigmatic pulses of high energy photons are sometimes seen in bursts lasting tens to hundreds of seconds, with rise times as short as a fraction of a second. Because the gamma ray detectors used until quite recently had a nearly isotropic response, source positions were determined by using the time delays between events at widely separated detectors. (Sound familiar?) Now, though, the most recent gamma ray burst detector, the Burst and Transient Source Experiment (BATSE) aboard the orbiting Compton Gamma Ray Observatory, is able to determine source positions all by itself (with errors of order 5 degrees) by comparing signal intensities among the various detectors on different sides of the spacecraft.

The earlier maps, based on rather crude positions, showed a featureless distribution consistent with complete isotropy. What is vexing about an isotropic map is that it might be produced by a distribution of sources at either of two vastly different distance scales. One possibility is that the sources are so close to the Earth (within several tens of parsecs, the thickness of the Milky Way's disk) that the overall flattened structure of our Galaxy, with its center far from the Earth in the direction of the constellation Sagittarius, does not appear. The other possibility is that the sources are so far away that their isotropy is indicative of the approximate uniformity of the distribution of matter on scales large compared to that of clusters of galaxies. Since a factor of 10^6 separates these distance scales, inability to distinguish these two possibilities leaves one with at least a factor of 10^{12} uncertainty in the luminosity of the source of the detected signals. It is hard to know how one could understand anything about a source of radiation without first finding a way to choose between these two alternatives. The study of gamma ray bursts has been crippled by just this dilemma.

There is some hope, though, that the Compton observations may provide the way to break the deadlock. The latest maps show a distribution with nearly perfect isotropy, extending over a broad range of signal strengths. The degree of isotropy is so remarkable, as is the fact that the isotropy has persisted as weaker signals have become visible, that a cosmological origin for the bursts is now becoming the preferred picture. Details of the event strength distribution, such as a deficit in the number of the weakest events, make sense as an evolutionary effect, if the sources were at cosmological rather than galactic distances.[207] One currently popular proposed emission mechanism is, in fact, based on the coalescence of neutron star binaries.[208] This means that gravitational wave observations, in particular the search for chirped gravitational wave signals whose end points coincide in time with gamma ray bursts, may help to resolve one of the most persistent mysteries of modern astrophysics.[209]

15.1.3 *Inferences from precise positions*

A wealth of additional information is available once one is able to determine positions to the few arcminute level. Once such a precise position has been determined for an individual source, it narrows down the list of potential electromagnetically visible counterparts to a few objects. Detailed observation may then reveal the unusual object. There is a vast array of tools available for investigating the nature of an object through its emission of electromagnetic radiation. Determination of positions precise enough to make this possible would be a great boon to gravitational wave astronomy.

One of the greatest astronomical discoveries of the second half of the twentieth century illustrates the point. Early radio telescopes had rather broad beam-widths, but as maps grew sharper, it became clear that many radio sources had a "double-lobe" morphology, often with a third more compact component between the first two. A major breakthrough occurred when Hazard and his co-workers made accurate measurements of the times at which the radio source 3C273 was occulted by the Moon. This showed, first of all, that the source consisted of two parts, one of which was nearly pointlike, and simultaneously determined its position to an accuracy of about 1 arcsec, coincident with a 13th magnitude blue star. This "radio star" had quite an unusual optical spectrum, however. Another compact radio source, 3C48, had also been identified with an unusual blue star by Sandage. The second shoe dropped when Maarten Schmidt recognized that the unidentified spectral lines in 3C273 made sense if the wavelengths had undergone a redshift $z \equiv \Delta\lambda/\lambda = 0.158$, a value unprecedented at the time. Interpreted as a shift due to the Hubble expansion, this proved that these "quasi-stellar" objects were more distant than any previously known object. (For more details of this story, see J. S. Hey's *The Evolution of Radio Astronomy*.[210])

Another episode, this time from the history of X-ray astronomy, carries a similar lesson.[211] Cygnus X-1, one of the brightest X-ray sources in the sky, couldn't be properly understood until an accurate position determined by a modulation collimator aboard a sounding rocket allowed radio astronomers to recognize a counterpart radio source, with a position determined to about 1 arcsec. This finally led to its identification with a particular blue supergiant star, HDE 226868. Subsequent optical observations determined the orbit of the star about the X-ray source — the orbital elements implied a mass for the latter object that was too large to be anything other than a black hole.

In both of these episodes (and in many more besides), basic understanding of the nature of a source of a new kind of radiation came about only after accurate positions allowed identification of an optical counterpart, and its subsequent intensive study by classical astronomical techniques. By now, there is nearly universal recognition of the virtues of multi-wavelength astronomy.

What will be required for gravitational wave astronomy to take advantage of this lesson of history? The crucial requirement is that the uncertainty in the position of the gravitational wave source be small enough that one can pick out with confidence an optical (or radio or X-ray) counterpart. The sky is thickly sprinkled with faint stars and distant galaxies, so it is by no means clear exactly what degree of precision will suffice to pick out the unusual object from the swarm of others. As long as our guess is true that the strongest frequent bursts will come from galaxies at roughly the distance of the Virgo Cluster, then a precision of tens of arcminutes may suffice. But if life turns out to be more complicated, then position errors not much larger than an arcminute may be required to make detailed links to the rest of astronomy.

Is that in the cards? Possibly, but it wouldn't come easily. Gürsel and Tinto examined the question at some length.[203] They showed that with a network of three detectors (two in the U.S., one in Europe) which could detect a burst with a signal-to-noise ratio of 45, one would achieve the required precision. How soon we will be in a position to detect gravitational waves with that SNR is not clear.

Another route to good position sensitivity would be to lengthen the baseline between detectors. If proposals to establish a permanent presence on the Moon come to fruition, one might very well think of installing a gravitational wave detector there. (The cost and difficulty of such a venture are of course highly uncertain.) It would give, in one direction anyway, a baseline roughly 30 times larger than is possible with terrestrial detectors. Assuming a lunar detector as sensitive as those on Earth, it would shrink the size of a position error box by a comparable factor.[212]

15.1.4 *Temporal coincidence with non-gravitational observations*

The strongly time-dependent signals we've been considering here offer another possible method for identifying an optical counterpart: look for something interesting that happens in the electromagnetic sky at the same time as the gravitational wave burst is received. This should in fact work well for the prototypical gravitational wave burst source, the Type II supernova. This is a phenomenon, after all, known for its optical display of remarkable brilliance.

During the season when the Virgo Cluster is visible at night, automated supernova searches now being inaugurated should reveal nearly every new supernova of

Type II within a day or two of its occurrence. Part of the delay might be due to the day-night cycle, part to the search pattern of the survey telescopes, and part due to the intrinsic time it takes for the star to brighten after the core collapse.

So, with a coincidence window of about a day starting with a detected gravitational wave burst, we can ask if an optical counterpart bursts into visibility. With an expected event rate of several per year, it might not be long after gravitational wave detectors of sufficient sensitivity go into operation before a joint optical-gravitational event demonstrated that strong gravitational waves were emitted in the core collapse leading to a supernova, and linked a gravitational wave signal to a specific event in a particular galaxy. This would allow a calibration of the gravitational luminosity of the core collapse.

This method could provide a powerful supplement to the direct methods of position determination discussed in the previous section. It is identical to the method by which the neutrino emission from Supernova 1987A was recognized and measured.[38] This occurred in spite of the fact that the neutrino detectors were not designed for neutrino astronomy, but instead to look for proton decay; furthermore, the detectors had only poor angular resolution. The occurrence of the neutrino signals in two detectors at a time just hours before the observed brightening of the supernova in the Large Magellanic Cloud was considered convincing proof of the connection of the phenomena.

Will this method actually be useful for gravitational wave astronomy? That depends on the gravitational luminosity of core collapse. a number which, as we saw in Chapter 3, unfortunately rather poorly known. But if such events emit as much as $10^{-3} M_\odot c^2$ in gravitational waves, then it seems quite likely that LIGO or another detector in its class will register gravitational waves from supernova events. And regardless of what is predicted before the measurements are carried out, any convincing temporal coincidences of gravitational waves and optical phenomena will give valuable insight into the physics of gravitational wave emission. There may also be a coincidence between a detection of a gravitational wave and cosmic neutrinos, if an optically "quiet" core collapse occurs nearby.

15.2   Interpretation of Gravitational Waveforms

We have been stressing the benefits of linking gravitational wave astronomy to other more traditional astronomical techniques. But it is important not to neglect the information that can be derived from interferometer signals themselves. If waveforms are measured with sufficient signal-to-noise ratio, we ought to be able to extract substantial insight about the sources of those waves. This will almost certainly give a view unavailable from other astronomical tools. As emphasized by Thorne,[9] the events emitting strong gravitational waves are typically not very amenable to observation by other means. Even the brilliant supernovae cloak the core collapses at their hearts in garments of nearly impenetrable opacity; the neutrino observations from SN1987A constitute the first direct observational evidence for the picture of stellar core collapse as the initiating event. Other events may emit strong gravitational waves without emitting any other form of radiation.

It is therefore necessary for us to be prepared to extract whatever information is available in the gravitational waveform itself. To learn the history of the rapidly changing mass distributions that emit strong gravitational waves, we need a means of making a faithful registration of a gravitational waveform. (Recall $h(t) \propto \ddot{I}(t)$.) To accomplish this requires a detector of sufficient sensitivity, and bandwidth spanning as many octaves as possible. These are really two aspects of the same problem. The continued effort to improve the performance of gravitational wave interferometers will be measured both by gains in overall sensitivity and by expanding the region of good sensitivity into a wider band of frequencies.

Assume for the moment that all of our hard work has paid off, leaving us in possession of a set of high signal-to-noise waveforms from a set of simultaneous observations by a worldwide network of gravitational wave detectors. How would we go about interpreting these data? We can best illustrate the process by discussing the key features of predicted waveforms from several possible sources.

We take as our examples three events that should be strong gravitational wave sources: the collapse of a rapidly rotating stellar core, the coalescence of a binary system of two neutron stars, and the emission of gravitational waves from the creation of a black hole or from the collision of two of them. What is striking is that the each of the waveforms carries a distinctive signature. If we obtain high SNR broad bandwidth measurements, we will have no difficulty in recognizing any one of these sources. That is an important realization, as it gives us some confidence that gravitational waveforms are interpretable. The examples also suggest how we might go about posing and answering questions about waveforms that weren't a good match to these particular cases.

15.2.1 *Core collapse*

Core collapses that lead to neutron star formation will generically contain two sorts of motion — free fall under the influence of gravity, and an abrupt *bounce* on a millisecond time scale governed by the repulsive forces between neutrons as the core reaches nuclear density. Some calculations suggest the possibility that the core may bounce several times, but each bounce should have comparable abruptness, separated by brief episodes of free fall by the core.

Each of the two sorts of motion results in a characteristic time dependence for the quadrupole moment of the core, and thus in a characteristic shape for the corresponding section of the waveform. So, if a generic collapse consists of some combinations of these two kinds of motion, then a generic collapse waveform should consist, at least roughly, of a combination of the corresponding waveform segments.

Another aspect of core collapse may give an additional distinctive signature. Unless the core angular momentum is greater than $\mathcal{J} \sim 10^{49}$ erg-sec, then we expect that it will remain an axisymmetric object throughout the collapse. The gravitational waveform emitted by an axisymmetric source is linearly polarized. We will be able to determine the polarization state of the received wave by comparison of the waveforms received at the various detectors in the global network. If the wave is linearly polarized, then each detector should give the same waveform shape, up to a multiplicative constant. The set of detected signals will be offset in time from each other by the travel time across the network for a wave travelling from the direction of the source. The amplitude ratios between sites will depend only on the polarization angle and on the position of the source.

On the other hand, a very rapidly rotating core will not remain axisymmetric. In this case the waveform will be, in general, elliptically polarized. (If the angular momentum vector is aligned with the line of sight, the polarization is circular, while if it is normal to the line of sight one recovers linear polarization.) This could be distinguished in the set of outputs from a network of detectors by a combination of a) non-matching waveforms, and b) shifts between waveform peaks not due to propagation delays across the network.

15.2.2 *Binary coalescences*

The coalescence of two compact objects will also have a distinctive signature that should be easy to recognize and interpret. Two neutron stars that spiral together due to the loss of energy through gravitational radiation follow an orbit that is calculable to a high degree of precision. The history of the resulting quadrupole moment is encoded in the elliptically polarized gravitational waveform.

A good approximation to the orbit was calculated by Clark and Eardley.[34] The resulting waveform has the form of a *chirp*, a quasi-sinusoid with frequency and amplitude both increasing functions of time, until the collision or destruction of the neutron stars. While the details of the waveform will depend on the masses of the stars and to a lesser extent on their spin angular momenta, the overall character of the waveform is a very robust indication of the nature of the source.

There is further information encoded in the waveform. The rate at which the signal sweeps up in frequency and strength is governed, to first order, by the single parameter $\mathcal{M} = \mu M_T^{2/3}$, where μ is the reduced mass and M_T is the total mass of the system.[213] While not offering an explicit opportunity to determine the stars' individual masses, it would at least allow a check of the likely hypothesis that more often than not a neutron star binary will consist of two objects each of about $1.4 M_\odot$, the Chandrasekhar mass.

It has recently been shown that, in order to predict the waveform to better than 1 radian throughout the chirp, one needs to include further corrections to the nearly Newtonian calculation. At this level, it turns out to be possible in fact to determine the individual masses of the two objects. There is the possibility of some confusion, though, if the neutron stars are spinning rapidly.[214]

The simplicity and predictability of the waveform does not last up to the very end of the event. In its last stages, the course of the interaction between the two objects depends on more than nearly-Newtonian gravity. In the case of neutron stars, the final events depend on the equation of state of nuclear matter. Tidal disruption is among the possibilities. If we are instead dealing with black holes, a fully general relativistic treatment of the final encounter is required. The wealth of possibilities makes the study of binary coalescence signals a wonderful laboratory for gravitational physics and astrophysics.

15.2.3 A gravitational standard candle

The distinctive signature and near-exact calculability of the chirp waveform in a binary coalescence suggests a remarkable possibility for gravitational wave astronomy. Schutz first noticed[213] that once one has extracted the mass parameter $\mathcal{M}$ from the rate at which the frequency changes with time, one is able to make a prediction for the binary's absolute gravitational luminosity. That means that a measurement of the amplitude of the signal measured on Earth is enough to determine the distance to the source.

In astronomical jargon, that makes a coalescing binary function as a *standard candle*. It would be unique in astronomy, for it would be the only standard candle whose luminosity can be calculated from first principles. And, as a standard candle perhaps visible across hundreds of Megaparsecs (assuming sufficiently sensitive interferometers), it offers a chance to behold the Holy Grail of modern cosmology, the value of the Hubble constant.

But determining the Hubble constant requires more than knowing the distance to an object, since it is defined as the ratio of a distant object's recession velocity to its distance away from us. It doesn't appear possible to also extract the redshift of the binary directly from gravitational wave observations. Instead, Schutz's method requires that an optical counterpart be found, so that its redshift can be determined by classical means. It isn't necessary to identify the specific galaxy in which the binary was located — identifying the cluster of galaxies to which it belongs is sufficient at the distances contemplated here. An angular error of order 1 degree is probably required.

This program ends up making strong, although not severe, demands on the angular resolution of the network, and hence on the signal-to-noise ratio of the detections. It is not certain that this is a truly practical alternative to more standard astronomical methods, which have, after all, already determined the Hubble constant to about a factor of 2.

The proposal does point out one crucial feature of gravitational wave astronomy. The signals will come from the coherent motions of large masses, and will correspond to a direct measure of those motions, the second derivative of the quadrupole moment $\ddot{I}$. Thus, if we attain a good enough SNR to record a signal well, there is a very good chance that we can understand it.

15.2.4   *Recognizing signals from black holes*

In the preceding discussion, we assumed for the sake of discussion that core collapse leads to the formation of a neutron star, and that binary systems of collapsed objects most likely would contain two neutron stars. These are quite reasonable situations to analyze, since we know of actual examples in the sky. However, in either situation, the collapsed objects might instead turn out to be black holes.

It is to the study of black holes that gravitational wave astronomy could make its greatest contribution. The quality of the match between the demands of the problem and the capability of gravitational wave detectors comes about, in essence, because a black hole is a phenomenon of pure gravitation. Once it has formed, it matters not what original constituents formed it. As a famous theorem states, "A black hole has no hair."[215] Its observable properties depend only on its mass, electric charge, and angular momentum. The link between the black hole and the rest of the universe occurs through the strong space-time curvature it produces in its vicinity. Operationally, what makes a black hole a black hole is the nature of that space-time curvature. Accretion disks observable by their X-ray emission or the rapid orbits of stars in nearby regions are only epiphenomena. What has so far been lacking is a method of observation that directly probes a black hole's gravitation.

With luck, the study of gravitational waves will be the suitable method. The reason is that whenever a black hole finds itself away from its equilibrium configuration, either by a small perturbation, a collision with another black hole, or in it original collapse, it relaxes to its quiescent state by vibrating in a characteristic set of highly damped *quasi-normal* modes.[216] These vibrations of the black hole horizon are in fact damped by their efficient emission of gravitational waves. So a black hole's most distinctive feature of "black holeness" is the emission of gravitational waves whose waveforms are calculable from general relativity directly.

An indicative black hole waveform was calculated by Stark and Piran.[217] A complete waveform begins with a *precursor* section that depends on the details of the disturbance that created or excited the black hole. But any black hole waveform ought to end with a tell-tale *ringdown* exhibiting one or a few of these modes, a distinctive signature unlike the waveform from neutron star gravitational collapse. Even if neutron star formation should occur with several bounces, it should be distinguishable by its free-fall sections separated by sharp reversals, as opposed to a smoothly damped sinusoid.

Detection of gravitational waves with the damped sinusoidal signature would thus indicate that the Universe does in fact contain objects possessing the essential nature that general relativity attributes to black holes. Furthermore, the frequency of the wave will allow us to read off the mass of the black hole M_{bh} directly — the gravest quasi-normal mode has a frequency $f_1 \approx 10$ kHz $(1 \; M_\odot / M_{bh})$.[9] More detailed study of the waveform would indicate the nature of the interaction between the black hole and its neighbors.

15.3 Previous Gravitational Wave Searches

The field of science known as gravitational wave astronomy is in some ways a strange one. It has existed as a working community of scientists for over two decades, or three if we count the '60s, when Weber worked alone. And yet, with the exception of Taylor's crucial but indirect result from the Binary Pulsar, there has not been a single confirmed detection of a gravitational wave. As experimenters have produced instruments of increasing sensitivity, they have paused occasionally to try those instruments out. Thus, the history of this period of the field is the story of a set of relatively brief "experiments", that have only set upper limits on the flux of gravitational waves reaching the Earth. The era of continuous observing of the sky is yet to begin.

15.3.1 *Room temperature bars*

The burst of activity following Weber's announcement of coincident pulses in two bars did not confirm the hope that he might have detected gravitational waves. On the contrary, no other groups observed any events that could be attributed to any influences external to the instruments.

To this day no straightforward explanation for the discrepancy has been found; an interesting insight into the nature of the debate in the early '70s can be found in the exchange of letters to the editor of *Physics Today* between Richard Garwin, a leading critic of Weber, and Weber himself.[218] A suspicion deeply held by many of Weber's critics was that he engaged in data selection and *a posteriori* statistical reasoning. Said one pair of critics,[219] "If one searches long enough in our finite sample of data, one must find some complicated property which distinguishes zero delay significantly from the others. (Again this is true for an arbitrary delay, but with a different property.)" The term "zero delay" refers to the statistical technique of calculating the number of coincidences between synchronized outputs of two bars as compared to the number that appear when an artificially introduced time delay is introduced between them. Discussion of the issue apparently reached an impasse when Weber's claim that others hadn't reproduced his apparatus faithfully enough was met with the arguments that (1) real gravitational waves ought to register in any similar apparatus of comparable sensitivity, and that (2) Weber hadn't revealed sufficient detail for truly identical instruments to be constructed.[220]

If we accept the inference that the irreproducibility of Weber's results means they should be put aside, then the null results of the other workers in the field set upper limits on the flux of kHz-band gravitational wave bursts. The most stringent upper limit from this era of experiments was that of the group at the Max Planck Institut (or MPI), then led by H. Billing.[219] They conducted a coincidence experiment between a bar located at Munich and another set up first at Frascati, later moved to Garching, a Munich suburb. Their result was expressed in terms of an upper limit on the rate of gravitational wave bursts, as a function of burst flux density. This limit appears as the rightmost curve in Figure 15.5.

15.3.2 *Cryogenic bars*

By 1981, the next generation of resonant mass gravitational wave detectors had begun to operate. The detectors in this class were cooled by liquid helium to a temperature of about 4 K. With the resultant reduction in thermal noise, and other improvements such as low noise superconducting transducers, the rms strain noise of these bars reached (and later went below) $h_{rms} \approx 10^{-18}$. The first example of this class to operate continuously was the bar at Stanford. Based just on the histogram of events in this single bar, Boughn *et al.* were able to set an upper limit on frequent burst events lower by about an order of magnitude in strain than the MPI coincidence experiment.[221] Their limit is also graphed in Figure 15.5.

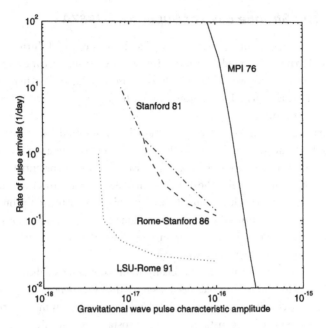

Figure 15.5: Upper limits set by various experiments on the rate of gravitational wave impulse arrivals, as a function of the characteristic amplitude of the impulse.

In 1986, the groups at Stanford, Louisiana State University, and Rome (whose bar was located at CERN in Geneva) carried out a three-way coincidence experiment among cryogenic bars.[177] This limit is shown as the third curve in Figure 15.5. The limit on gravitational wave burst flux is only slightly improved, since all three bars were noisier than they ought to have been. The best limit came from two-way coincidence between the Rome and Stanford bars, because the LSU detector was not performing as well as the others.

By 1991, the problems in the LSU bar had been corrected. In fact, it set a new record (which still stands) for rms sensitivity to kHz band bursts, $h_{rms} \approx 6 \times 10^{-19}$.[222] Subsequently, a new coincidence run with the Rome group was carried out. (The Stanford 4 K detector was irretrievably damaged in the Loma Prieta earthquake of 1989.) The upper limit derived from this run,[223] represents the most stringent yet set; it is shown as the lowest curve in Figure 15.5.

Development continues in cryogenic bar technology. A new generation of ultra-cryogenic bars, cooled by helium dilution refrigerators to temperatures of about 50 mK, is about to come on line. As of this writing, late 1993, a trial cool-down has been carried out at Rome, and Stanford is not far behind.

15.3.3 *The Strange case of Supernova 1987A*

The 1986 coincidence run of the three cryogenic bars occured during the period April–July of that year. Immediately thereafter, all three groups warmed up their bars for further development work. As a result, on February 23, 1987, when Supernova 1987A occurred (as seen on Earth), none of the world-class detectors of gravitational waves was "on the air".

In fairness, it is unlikely that gravitational waves would have been detected. Optimistic best guesses put the strength of a wave burst from a supernova 10 kpc away, typical of distances in our galaxy, at $h \sim 10^{-18}$. This is substantially below the level of the non-Gaussian tail of the distribution of events in the three bars during the run. Even if the bars had shown only Gaussian noise, the detection probability for an event of that strength would have been marginal. Since the supernova occurred in the Large Magellanic Cloud, at a distance of about 50 kpc, even optimists would have been surprised if a detection were made.

Indeed, they were surprised at the claims made soon afterwards that something associated with the supernova had been seen by the two room temperature gravitational wave detectors, at Rome and Maryland, that had been in operation at the time. Without the need to constantly refill a cryostat, and with development work devoted to new detectors, it was an inexpensive and scientifically sound strategy to monitor and record the outputs of these relatively insensitive but robust instruments. Shortly after the supernova's appearance, word spread that it had been seen in these instruments.

The detailed story[224] is rather complicated, and has been greeted by intense skepticism by most of the rest of the gravitational wave community.[225] The claim of the Rome and Maryland groups is that there was a two hour period of excess noise in their detectors that coincided with the time of the supernova, as seen in the Mont Blanc neutrino detector. More specifically, if one constructs a template out of the times at which individual candidate neutrinos were registered in the Mont Blanc detector in a two hour window centered on the supernova time, then looks at the outputs of the Rome and Maryland bars at those times, the outputs were anomalously high. There is some additional "signal" if one expands the time window further. The statistical significance claimed for the anomaly is roughly at the part in a thousand level.

Reasons to be skeptical are numerous. For one thing, most of the neutrino physics community puts the time of the event some four hours later than the Mont Blanc team, at a time when coincident neutrino bursts were registered in the 1 MB and Kamiokande detectors. Secondly, the signature of the event in the bars is not of the sort that had been predicted in advance, a single millisecond pulse (perhaps with some structure on that time scale) marking the moment of gravitational collapse. The significance of the result depends on the two hour width of the window, whose connection with the supernova event is dubious at best. Add this to the physical implausibility of gravitational waves of sufficient strength to register in such insensitive devices, and it is no wonder that few have been convinced of the claims. Even the authors themselves admit that something is very strange. One of them summed up the situation in the following words[226]:

Our experimental result is far from being understood mainly for the following reasons:

a. The g.w. signals are larger by a factor of at least 10^5 than those expected if the cause was gravitational waves originating in the Supernova.
b. The neutrino detection is stochastic in nature, not deterministic, and therefore we expect, if several ν bursts arrive at the detector, that sometimes the detector sees just one ν, other times two, three, zero, etc. according to the Poisson law. In our case, instead, the individual neutrinos appear to be preferred.
c. The collapse activity lasts at least 5 hours.

Point (a) is certainly the most difficult to understand. If one cannot use a larger antenna cross section, one is forced to consider gravity theories alternative to GR or to go towards new physics: i.e. new particles or unexpected behaviour of cosmic space. Point (b) might require abandoning the idea that individual neutrinos are detected. After all, the used instrumentation does not reveal the tracks of the individual interactions in all the cases.

(The mention of "a larger antenna cross section" is a reference to an alternative derivation of the interaction of a bar with a gravitational wave, due to Weber.[227] If true, then resonant mass detectors would be much more sensitive than had previously been believed. The logic of the derivation, and its result, have been strongly criticized by, among others, Grishchuk.[228])

Again, this strange claim may have resulted from the use of *a posteriori* statistical reasoning.[225] It is a slippery thing to calculate how unlikely an event is, if the signature of the event is not decided until after the data is examined for unusual features. To be sure, the authors of this purported detection are aware of this danger, and have tried to check the original discovery by looking for unusual features in the outputs of other neutrino detectors. But a close reading of their later analysis reveals many cases of "tuning" the signature to continually maximize the strangeness of the result. It has, however, been difficult to demonstrate that such tuning accounts for the entire statistical significance of the anomaly.

Perhaps the soundest lesson that can be drawn from the whole affair is that gravitational wave astronomy will not truly come into being as a science until a number of events are detected unambiguously. Too much concentration on one singular event is almost never a good policy. This is the deepest motivation to continue to improve the sensitivity of gravitational wave detectors — at sensitivity of 10^{-21} to 10^{-22}, there is hope that events can be detected once a month or more, not the once in thirty years associated with Milky Way supernovae. This continues to be the goal of the entire gravitational wave community, both the bar groups and interferometer designers; it drives the continuing improvement program of resonant mass detectors, and provides the fundamental design rationale for LIGO and the other large interferometer projects.

15.3.4 *Gravitational wave searches with interferometers*

Compared to bars, there has been less emphasis on using the interferometers that have been built so far to look for gravitational wave signals. This is primarily due to the fact that present-day devices are considered primarily to be laboratory prototypes for the large interferometers like LIGO. Lack of motivation to take a pause in development work to search for signals may also come from the fact that these small prototypes have never yet held the lead over bars in gravitational wave sensitivity, although they do not lag far behind.

Nevertheless, there have been some trial uses of interferometers as data gathering instruments. Hereld used the Caltech 40-m interferometer to search for periodic signals from the millisecond pulsar, 1937+21. He set 3σ upper limits of 1.1 to 1.5×10^{-17} on the strength of the gravitational waves at the second harmonic of the pulsar rotation rate, and comparable limits on the signal at the pulsar's rotation frequency.[229] Dewey used data from the 1.5-m interferometer prototype at MIT to set a limit on the flux of a variety of impulsive signals. Using only the data from this single instrument, he was unable to reject as possible signals some events as large as $h \sim 5 \times 10^{-14}$.[159] Livas, using data from the same instrument, looked for any periodic gravitational waves in the frequency band 2–5 kHz. An all-sky survey, making no Doppler corrections, yielded a 5σ upper limit of $h_{rms} \approx 5 \times 10^{-17} (2000 \, \mathrm{Hz}/f)^2$ in that band. Applying Doppler corrections appropriate to a source at the Galactic Center, Livas set a 7σ upper limit of $h_{rms} \approx 2 \times 10^{-17} (2000 \, \mathrm{Hz}/f)^2$ on any periodic signals from that location.[196] Smith used data from the 40-meter interferometer to search for waves with the chirping signature of neutron star binary coalescence. Her upper limit on event strength is 5×10^{-17}.[230] Zucker studied four samples of

105 seconds worth of data from the Caltech 40-meter interferometer, in a search for periodic signals from a possible pulsar at the location of Supernova 1987A. He was able to set a limit of $h_{rms} \approx 1.2 \times 10^{-18}$ on the strength of waves from that location, in the band between 305 Hz and 5 kHz. He also used the same data to rule out periodic signals stronger than 6.2×10^{-19} from the direction of the galactic center, in the band between 1.5 kHz and 4 kHz.[146]

The most serious attempt at observation with gravitational wave interferometers to date was a coincident run of the Max Planck Institute's Garching 30-meter and the University of Glasgow's 10-meter interferometers for a 100 hour period spanning March 2–6, 1989. The arrangements for this run were made in something of a hurry, so neither interferometer could be adjusted to its optimum sensitivity. Nevertheless, it marks a milestone as a true simultaneous trial run of interferometers with good performance. The results answered worries about the ability of such complex instruments, with their many servo loops, to operate reliably for long periods. The Garching interferometer achieved an in-lock duty cycle of about 99%, with little manual intervention. The Glasgow interferometer also achieved a very high in-lock duty cycle.[232]

Analysis of the data from the 100 hour run has shown that the probability distributions of the time series from the two interferometers are accurately Gaussian out to at least $4\,\sigma$, once the data have been edited to remove segments when the rms noise was unusually large. Then, an upper limit of 3×10^{-16} can be set on the size of frequent gravitational wave bursts.[233] The Garching data have also been analyzed to look for possible gravitational waves from a pulsar at the location of Supernova 1987A. This analysis was specialized to bands 4 Hz wide, centered on the fundamental frequency and second harmonic of the optical pulsar reported by Kristian et al.,[234] and subsequently retracted by Kristian.[235] An upper limit of $h \approx 9 \times 10^{-21}$ was set on the amplitude of periodic gravitational waves from that spot, in those bands.[232]

15.3.5 *Other observational upper limits*

There are other methods, besides resonant bars and interferometers, that have been used to search for gravitational waves. It would take us too far afield to discuss these methods in detail, so we will only give a brief overview, but urge the reader to consult the references cited below.

Interplanetary spacecraft have been used as test masses in gravitational wave observations. A radio signal whose frequency is locked to that of a high quality atomic clock is transmitted from a ground station to the spacecraft. On board, a transponder retransmits a new signal that is locked in phase to the received signal;

this returned signal is recorded back on Earth, where its frequency is compared with that of the master oscillator. The principle of the system is analogous to a one-armed interferometer, in that the radio wave is used as an electromagnetic length standard to measure the separation between two freely falling test masses, the Earth and the spacecraft. Of course, the two test masses are not at rest with respect to each other, so the measurement is more conveniently done by looking for fluctuations in the frequency of the returned Doppler-shifted signal. So far, the experiments have involved only one spacecraft at a time, but two or more could be used to give rejection of clock frequency noise. One of the other most important noise sources in this measurement is fluctuations in the index of refraction of the interplanetary medium. Measurements to date are most sensitive at gravitational wave signal frequencies between 1 and 10 mHz. Analysis of data from the Voyager spacecraft yielded upper limits of $h \leq 3 \times 10^{-14}$ at 1 mHz.[236] The Galileo spacecraft was expected to yield sensitivity an order of magnitude better, because both *uplink* and *downlink* were carried out at X-band (8.4 GHz) frequencies; Voyager used an S-hand (2.1 GHz) uplink, in which the solar wind noise is stronger. A review of this branch of the field is provided by Hellings, in his chapter in Blair's *The Detection of Gravitational Waves*.[237]

Observations of naturally occuring systems also have been used to look for gravitational waves. The excitation of quadrupolar modes of oscillation in the Earth or the Sun ought to be a probe of the amplitude of gravitational waves at the resonant frequencies of those modes. In effect, these bodies are being used as very large, although not especially quiet, resonant mass detectors. The modes in the Earth can be measured with long period seismometers. (The fundamental quadrupolar mode has a period of 53.9 minutes.) The modes of the Sun are observed with the remarkably successful Doppler shift methods that have created the field of *helioseismology*. In neither case are the modes ordinarily visible above the noise; the Earth's quadrupolar mode has been observed only after large earthquakes. The most careful interpretation of the available data is probably that of Boughn and Kuhn.[238]

The fact that some pulsars function as ultrastable clocks has also been used in the search for gravitational waves. This observation implies an upper limit on the strength of gravitational waves traversing the intervening space; were they stronger, they would make the pulse train fluctuate in its arrival rate. Again, we have an analog of a one-armed interferometer, although this one works by comparing two different clocks, one at each end. Observations of several stable pulsars allow one in principle to make a discovery, not just to set an upper limit. The best limit to date by this method was set by observations of four different quiet pulsars.[239] At a frequency of 10^{-8} Hz, the upper limit on the strength of a stochastic background of gravitational waves is $h(f) \approx 3 \times 10^{-8}/\sqrt{\text{Hz}}$, in a band of order an octave wide. This may not sound like a particularly impressive upper limit when expressed in these terms, but it corresponds to a restriction that the energy density in such a background cannot exceed 1 part in 10^3 of the density needed to close the Universe. A particularly good clock is provided by the pulses of the Millisecond Pulsar (PSR 1937+21); on time scales of six months or longer, they are spaced with a regularity as strict as that of the best atomic clocks on Earth.[240] This leads to a limit that a gravitational wave background near 10^{-7} Hz can have no more than 5×10^{-4} of the cosmic closure density.

Even astronomical observations on the largest scales can say something about the amplitude of a possible gravitational wave background. The degree of isotropy of the cosmic microwave background radiation would be disturbed if very long wavelength gravtitational waves filled the Universe. In effect, we use the temperature of the background as our length standard. Interpretation of pre-COBE data was discussed by Carr.[59] The map from COBE's DMR experiment, showing departures from perfect isotropy, is now being scanned for traces of a gravitational wave signature.[241]

Prospects   **16**

Encouraged by the prospects for astrophysically interesting sensitivity to gravitational waves, research groups around the world have been contructing laboratory scale interferometers since the late 1970's. Several, including those at MPQ-Garching,[79] the University of Glasgow,[242] and Caltech,[6] have achieved rms sensitivities nearly rivalling the best cryogenic bar detectors. The main aim of this work, though, has been the development of techniques that will make possible the successful operation of interferometers with arm lengths of 3 to 4 km.

16.1 A Prototype Interferometer

An interferometer with arms of $L = 40$ m has operated at Caltech as a test bed for the 4 km LIGO interferometers.[6] It has been arranged, up to now, in the "simple Fabry-Perot" configuration discussed in Section 12.6 above. That is, each arm length is sensed separately without recombination of the light at the main beam splitter. Ultra-high reflectivity mirrors allow sufficient storage time τ_{stor} for the sensitivity to be optimum for signals of frequency $f > 300$ Hz. A highly stabilized argon ion laser supplies up to 5 Watts of light to the system, although various correctable losses limit the useful power to of order a tenth of a Watt. The 3.7 cm diameter mirrors are optically contacted to 1.6 kg fused silica test masses. Those masses are suspended as 1 Hz pendulums from multi-layer lead-rubber isolation stacks. The whole system is evacuated to a pressure of 2×10^{-5} torr.

As of June 1992, the noise spectrum of the 40 m interferometer was that shown in Figure 16.1. (The interferometer was subsequently shut down for a major upgrade, and has only recently come back "on the air".) Above 1 kHz, the spectrum is determined by shot noise. Below about 200 Hz, the noise is primarily caused by

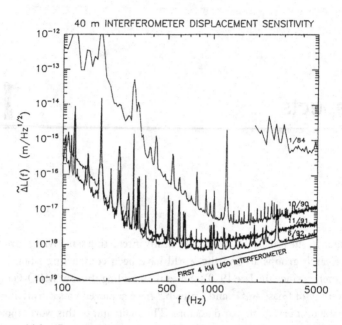

Figure 16.1: Recent noise spectrum of the 40 meter interferometer at Caltech.

seismic noise driving the isolation and suspension system. The intermediate region includes resonant peaks driven by a combination of seismic noise and thermal noise.

The broad-band noise is low enough in the 500 Hz to 1 kHz band that the rms noise in a search for 2 msec bursts would be about $h_{rms} \sim 10^{-18}$.

16.2 LIGO

The U.S. National Science Foundation recently began funding the construction of a pair of $L = 4$ km interferometers, a project that goes by the name of LIGO (for Laser Interferometer Gravitational-Wave Observatory). It is implemented by a team of researchers from Caltech and MIT. Once operational, LIGO will function as a national facility for gravitational wave research. A review of the key features of the LIGO project may be found in a recent article by the members of the LIGO Science Team.[6]

LIGO will consist of a pair of interferometer facilities at widely separated locations within the United States, managed as a single gravitational wave observing system. Livingston, Louisiana and Hanford, Washington were recently designated as the LIGO sites. At each location, a vacuum system will be constructed for interferometers with $L = 4$ km. The beam pipes will have a clear aperture of 1.0 meter, and will be evacuated to pressures of 10^{-9} torr. This would allow installation of interferometers with delay-line folding, but Fabry-Perot arms have been chosen instead; they will allow the large pipes to enclose several interferometers side by side. Initial plans call for one of the two sites to have an $L = 2$ km interferometer operating in parallel with the full length interferometer, for use in the manner discussed in Section 14.5. Eventually, each site would be equipped with several interferometers. Clever vacuum chamber design ought to allow construction and modification of interferometers with only negligible mutual interference.

Construction funding for the LIGO project was approved in Fiscal Year 1992. The VIRGO project, an Italian-French collaboration to construct one $L = 3$ km interferometer, was approved in 1993; it will probably be built on approximately the same schedule as LIGO.

16.3 Proposed Features of 4 km Interferometers

Plans for the first generation LIGO interferometers are at an advanced stage of development. Its key features have been sketched as:

- recombined Fabry-Perot optics,
- 5 W argon ion laser as light source,
- recycling of "bright fringe" light, increasing effective input power by a factor of 30,
- fused silica monolithic test mass mirrors of 10 kg mass,
- 1 Hz pendulum suspensions with $Q \approx 10^7$,
- multi-stage passive isolation stacks.

The noise power spectrum of this instrument, scheduled to go "on the air" around 1998 at both LIGO sites, should resemble that shown in Figure 16.2. The limiting noise sources should be: seismic noise below $f \sim 60$ Hz, shot noise above $f \sim 200$ Hz, and thermal noise due to the test mass internal modes between $f \sim 60$ and $f \sim 200$ Hz, assuming those modes have quality factors of order 10^6. With this noise spectrum, the rms noise in a search for several-msec bursts should be $h_{rms} \sim 3$ to 5×10^{-22}.

Is this good enough to guarantee successful detection of the sources we discussed in Chapter 3? Not quite, but it is definitely in the range where detection of gravitational waves would hardly surprise us. It represents an improvement of roughly three orders of magnitude in amplitude sensitivity over the best searches yet conducted. Being able to operate a search at this level represents a major advance with good prospects for success.

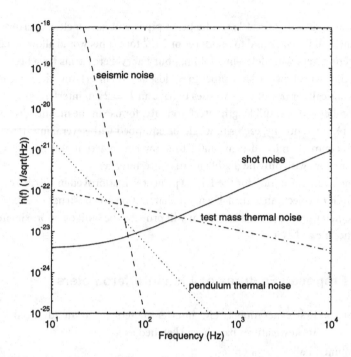

Figure 16.2: An estimate of the strain noise spectral density in an early LIGO interferometer.

Nor is the performance of this planned interferometer the best we expect to be able to achieve. As good as this first generation instrument will be, it is a deliberately conservative extrapolation of interferometer technology. All of the limiting noise terms are susceptible to substantial improvements, without modification of the capital facilities (vacuum system, beam pipes, laboratory buildings, etc.) Shot noise may be lowered by application of higher power lasers, more ambitious recycling, and appropriate application of signal recycling techniques. Seismic noise may be drastically lowered by construction of an isolation stack with lower resonant frequencies (like the Pisa Super Attenuator), or by utilizing one of the active schemes that we discussed in Chapter 11. Thermal noise, too, ought to be reducible by selection of lower loss suspension materials.

None of these improvements will come without further hard work on the appropriate subsystem. But the potential performance improvements are substantial. Reductions of shot noise by an order of magnitude or more are possible. More aggressive seismic isolation may move that "wall" down in frequency by about a decade. If thermal noise can also be substantially reduced, then we would have a detector that would be almost certain to detect cosmic gravitational waves. Coalescences of neutron star binaries ought to give frequent signals with amplitude of several times 10^{-22}. These signals ought to be detectable in a second generation interferometer.

Epilogue   17

17.1 Introduction

As I write this epilogue in the summer of 2016, twenty-two years have passed since the publication of the first edition of *Fundamentals of Interferometric Gravitational Wave Detectors*. (That is the same amount of time that separated *Fundamentals of Interferometric Gravitational Wave Detectors* from Rai Weiss's 1972 prospectus for LIGO in the *MIT Research Laboratory of Electronics Quarterly Progress Report*.[243])

More important than simply the passage of twenty-two years, those years have seen the realization of large interferometers, especially LIGO and Virgo but also the smaller but still crucially-important GEO600. If I had written one year earlier than this, I would have been talking about the technical triumphs of our instruments, and about the coming dawn of the Advanced Detector Era in which a factor of ten improvement of sensitivity might be the boost that the field needs to start detecting signals.

But since I am writing after 2015 September 14 09:50:45 UTC, the story that I have to tell is of the astounding success of this field; the successful implementation of technology certainly, but most essentially the detection of a gravitational wave signal from a black hole binary system with masses of 36 and 29 solar masses.[244] GW150914 was found at the very beginning of the first observing run of Advanced LIGO, with sensitivity half an order of magnitude better than that of the initial interferometers, but still half an order of magnitude shy of Advanced LIGO's design performance.[245] LIGO's O1 observing run also revealed a second binary black hole signal, GW151226.[246]

If you are a reader of this new edition, drawn to the field by the excitement of the new discoveries, you probably are looking for guidance on how to use a book written decades earlier to learn what you need to know about interferometric gravitational wave detectors. That is what I would like to offer you here.

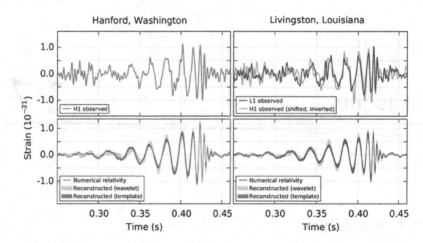

Figure 17.1: Filtered time series of the waveform of GW150914. See Reference 244 for details.

This book was intended as an entry point for someone who wanted help in learning the basic physical principles behind gravitational wave detection with interferometers. Those principles are indeed what underlies Advanced LIGO and Advanced Virgo. If you want to understand:

- how a gravitational wave interacts with an interferometer and generates a signal, or
- how thermal noise or seismic noise might obscure that signal, or
- how servo systems help to tame an otherwise twitchy detector and make it linear over a broad dynamic range, or
- how matched filtering lets one find a weak signal buried in noise,

you will find a path to that understanding in this book.

Since 1994, several other excellent books have been published that are well worth careful study. Readers should seek out books by Maggiore[247] and by Creighton and Anderson,[248] which give introductory takes on the field with somewhat different emphases.

At the more advanced level, there are volumes edited by Blair *et al.*[249] and by Bassan.[250] I should also mention that David Reitze and I are in the final stages of editing what we hope will be a definitive summary of the technology of the Advanced Detector Era, *Advanced Gravitational Wave Interferometric Detectors.*[251] If you are looking for the technical details of the detectors that are opening the field of gravitational wave astronomy, these will be the books for you.

There is a story in the Talmud[252] of the gentile who came to Rabbi Hillel and asked to be taught the essence of Judaism while he stood on one foot. Hillel taught him the Golden Rule, and then said, "The rest is commentary, go and learn it." I am under no illusions that the question of how to detect gravitational waves is as important as the question of how to lead a good life. I can say, though, that having mastered this book will still leave you with a lot of "commentary" to learn.

The remainder of this epilogue can be thought of as a program of study for a student who has mastered the basic physics and engineering ideas, as taught in the previous chapters. Here is a guide to all of the rest of what you need to know. Go and learn it.

17.2 Physics/Engineering Background (Chapters 4, 10, 11)

As a summary of essential foundational ideas, I can still recommend that a beginner make an intensive study of Chapters 4, 10 and 11. Chapter 4 contains very standard "applied mathematics for experiments"; the main virtue that I would claim for this chapter is that it pulls together, in a concise way, a set of useful topics that aren't often found in the standard physics curriculum (and that are scattered across the engineering curriculum). Chapter 11 on control systems aspires to be a mini-course on an absolutely essential topic that is standard in the mechanical engineering curriculum but that students of physics are expected to teach themselves. We physicists should claim this material for ourselves; servos are extremely useful, technically sweet, and profoundly beautiful.

Chapter 10 treats the concept of a null instrument, a topic that isn't formally taught anywhere, but that forms one of the core principles of experiment design. The gist is this: The cleanest way to measure a weak effect is almost always to hold the instrument at a null output, and to find the signal in the strength of the feedback required to hold the instrument at the null. To avoid the nearly-universal excess noise that afflicts measurements at low frequencies, one almost always applies a small high frequency modulation about the null output, and looks for a signal to appear at the modulation frequency.

The classic but obscure paper (in experimental physics, those characteristics seem to go together) by Roll, Krotkov, and Dicke[253] remains the best place to learn this material "in the original". Whatever Rai Weiss hadn't already learned from his first mentor Jerrold Zacharias, he had a chance to learn at the master's knee when he was a postdoc at Princeton with Dicke in 1962–64. Weiss's own classic-but-obscure

1972 paper shows how thoroughly he had learned the lessons of designing a null instrument.

Given this, it is interesting that a genuine novelty in null instrument design became a key feature of the Advanced LIGO interferometer. The arm length difference of the interferometer (the key degree of freedom that senses the presence of a gravitational wave) isn't modulated at all! Instead, using a system called "DC readout", the operating point of the interferometer is maintained at a small steady bias away from the null output, with the signal being read out directly in the audio band.

Many of us trained in the school of Dicke and Weiss were astonished by this innovation. The short explanation goes like this: Almost every other degree of freedom in a LIGO interferometer is measured by RF modulation and held at its operating point by a null servo. By the time one comes to interrogate the arm length difference, the DC effects that would make technical noise in this degree of freedom have been dramatically reduced (in large part by the low pass filtering inherent to the optical system itself), eliminating the need for high frequency modulation. A more careful explanation can be found in the paper by Fricke et al.[254]

17.3   Prehistory of Gravitational Wave Detection (Chapter 1)

Readers who would like to know more about the prehistory of gravitational wave detection should study Daniel Kennefick's essential history, *Traveling at the Speed of Thought: Einstein and the Quest for Gravitational Waves*.[255] It explains how it could be the case that Einstein and many other physicists doubted the physical reality of gravitational waves for the first four decades after Einstein's initial prediction of their existence. It also gives an excellent account of the Chapel Hill Conference of 1957, at which those doubts were finally and completely dispelled.

I would also like to refer readers to an article[256] that I wrote paying tribute to my Syracuse colleague Josh Goldberg who, in his role in 1957 funding gravitational research for the United States Air Force, was the sponsor of the Chapel Hill conference. In the article, I emphasize the importance of the late Felix Pirani's contribution to the resolution of the question of the reality of gravitational waves at the conference. I also argue that Joseph Weber's quest to detect gravitational waves began at the exact moment that he heard Pirani's talk in Chapel Hill. Thus, we can date the start of the effort to detect gravitational waves to the week of January 18–23, 1957.

For more recent history, readers are urged to seek out Marcia Bartusiak's classic *Einstein's Unfinished Symphony*[257] (soon to be available in an updated edition). A very up-to-date account of LIGO's early history is given by Janna Levin in *Black Hole Blues*.[258] Readers interested in understanding how LIGO's story informs our understanding of science should study Harry Collins's magisterial *Gravity's Shadow*[259]; the story continues in *Gravity's Ghost and Big Dog*,[260] and will conclude with the forthcoming *Gravity's Kiss*.[261]

17.4 Gravitational Waves and their Interactions with Detectors (Chapter 2)

The account given in Section 2.3 of how gravitational waves cause a change in the output of a Michelson interferometer was based on Weiss's 1972 implementation of Pirani's prescription (as elaborated in the so-called "Blue Book" of 1983).[262] This treatment has stood the test of time. You will not go wrong if you focus on how gravitational waves change the light travel time between freely-falling test masses.

I offered a heuristic interpretation of this calculation in a paper of 1997, "If light waves are stretched by gravitational waves, how can we use light as a ruler to detect gravitational waves?"[263] My intention was to show how the standard correct calculation could be understood in terms of experimental intuition. On a topic as subtle as the physical interpretation of general relativity, heuristics are tricky (and perhaps even dangerous). This one in particular has prompted a lot of commentary. For a very clear and especially gracious critique (so gracious that it doesn't mention the paper that it is criticizing!), see the article by David Garfinkle.[264]

17.5 Sources of Gravitational Waves (Chapter 3)

As an account of basic physics, Chapter 3's very brief summary of gravitational wave sources is still useful. But this is a subject on which, thankfully, there's been a tremendous amount of progress. Today's reader should surely know a lot more.

The discovery of GW150914 makes it seem prescient that binaries were the featured source in my 1994 treatment, but I like many others thought that neutron star binaries were the most reliably predicted sources (true enough) and also the most likely to be found first (proven false). The summary of Peter Kafka's way of framing the amplitude of the signal from a binary[265] is more relevant than ever — GW150914 had a maximum amplitude of 1×10^{-21}, with masses of about 30 solar masses and a distance of about 30 times the distance to the Virgo Cluster.

In the past decade, the most dramatic progress in our understanding of gravitational wave sources came from the success of numerical relativity. Starting with the paper of Frans Pretorius in 2005[266] followed immediately by results from the RIT[267] and the NASA-Goddard groups[268] and never slowing down since then, the calculation of the expected waveform from a black hole binary became a solved problem (at least at low spin and modest mass ratio). The importance of that work for the discovery and interpretation of GW150914 should not be understated. There is still important work being done on the case of black holes with large misaligned spins and extreme mass ratios; meanwhile, other workers in the field have moved on to the richer problem of binaries containing neutron stars or supermassive black holes in the complex environment of galactic nuclei. Much more sophisticated work has also been carried out in recent years on calculating the expected abundances of binaries and thus the consequent event rates. An "official" LSC-Virgo summary of the literature is given in Reference 269; the error bars in those estimates spanned several orders of magnitude. Now that black hole binary signals are being detected, those estimates are in the process of being replaced by measurements.[270]

Calculations of the gravitational wave signals from core-collapse supernovae have also made tremendous progress since the first edition of this book.[271] As the physical treatment of the problem has become much more sophisticated, optimism about the possible strength of these sources has receded, but there will be much to learn once these signals are eventually detected.

Much more is also known about possible sources of continuous wave and stochastic signals — see the references to papers on those searches, below.

17.6 Quantum Measurement Noise (Chapter 5)

The separate treatment of shot noise and radiation pressure noise presented in Chapter 5 now seems quite dated, although some readers may want to hold onto that treatment for its heuristic value alongside the more correct (but more subtle) treatments available today.[272] For the design of the advanced generation of interferometers, it was essential to use the correct theory. In fact, if thermal noise from mirror coatings had not turned out to be as strong as it (currently) is, details of the interplay between the shot noise and radiation pressure aspects of quantum measurement noise would soon be dominating the sensitivity of interferometers in their most sensitive frequency band.

Shot noise has been driven to remarkably low levels; fractional measurements of order 10^{-10} of a fringe are called for in advanced interferometer designs. Nd:YAG lasers with output powers of up to 200 W are parts of the Advanced LIGO and Advanced Virgo designs. To tame the radiation pressure noise, mirrors for advanced interferometers were increased in mass to 40 kg (from the 10 kg of the initial generation).

Figure 17.2: A pre-stabilized laser for Advanced LIGO, undergoing final adjustments. (Credit: Caltech/MIT/LIGO Lab).

Remarkably, squeezed light has developed very rapidly. It is now used routinely in the GEO600 interferometer, where it reduces shot noise by more than 3 dB.[273] An experimental demonstration of the use of squeezed light was carried out on one of the LIGO interferometers.[274] While not a part of the baseline design for Advanced LIGO or Advanced Virgo, it might be one of the first upgrades to the design.

(I'd also like to point the reader to a more careful derivation of the shot noise limit by Lyons *et al.*[275])

17.7 Interferometer Configurations (Chapters 6 and 12)

Fabry-Perot cavities are used in almost all of the current interferometers. (GEO600 is the sole exception.) Thus, the material on Herriott delay lines in Section 6.1 now seems quaint. Power recycling was an essential part of all large interferometers in their first generation, and signal recycling is equally so for advanced interferometers.

An excellent account of how all of this comes together (and what additional features were required) can be found in the overview articles on initial LIGO,[276] Virgo,[277] Advanced LIGO,[278] and Advanced Virgo.[279]

17.8 Thermal Noise (Chapter 7)

Thermal noise has proven to be a very important noise source in the mid-frequency band. While the basic physics as described in Chapter 7 has held up, much has been

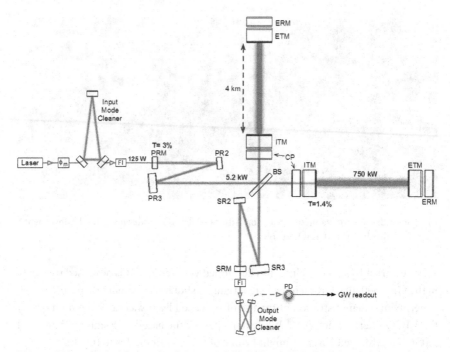

Figure 17.3: Optical layout for Advanced LIGO. For additional information, see Reference 278.

learned, leading to a rather different understanding of <u>which</u> aspects of thermal noise are most important in interferometers.

The near-ubiquity of constant-loss-angle internal friction is such that it now has a nickname: structural damping. A nice demonstration of the impact of structural damping on the thermal noise spectrum of a "toy" system was made by González.[280] Since then, special purpose instruments have measured the thermal noise of realistic mirrors.[281,282]

One place where Chapter 7 really missed the mark was in the detailed treatment of the most important form of thermal noise, that coming from thermal fluctuations of the mirrors' surfaces. Several effects are now understood that weren't understood in 1994. The notion that thermal noise is dominated by the contribution of the gravest one or two modes was shown to be seriously in error by Gillespie and Raab[283]; the sum over all modes turns out, unfortunately, to lead to a substantially larger noise level. Levin[284] showed an elegant way to understand this result, which also made it clear that the internal friction in mirror coatings might play a dominant role because that dissipation is applied right where the interferometer's light interacts with the mirror. Subsequent experiments[285, 286] proved that internal friction in coatings was high enough that Levin's worry came true.

There was some good news, too. Fused silica was shown to be an outstandingly low-loss material. In the advanced interferometers, fused silica fibers make the pendulum-mode thermal noise small enough that Newtonian gravitational noise competes with it in strength; the necessary technological breakthrough in the assembly of all-fused-silica suspensions came with the invention of the hydroxide-catalysis bonding technique, developed for the Gravity Probe B satellite.[287] Meanwhile, best-quality fused silica, properly fabricated and with proper heat treatment, yields a mirror where the substrate thermal noise is negligible compared with the coating noise.[288] If only we could make better coatings! The book by Harry et al. is a good recent review of the state of the art in coatings.[289]

17.9   Seismic Noise (Chapter 8 and Section 11.7.2)

Vibration isolation has played as big a role in interferometer design as was expected. As with so much else in the field, the basic physics is as it was described in 1994, but the technology has developed quite far. The most advanced (and well-engineered) isolation system at that time was the Virgo Super Attenuator. The version described in Chapter 8 used air springs for vertical isolation. Eventually, Virgo settled on a much better solution, involving pre-stressed cantilever blade springs made of a special alloy, maraging steel. [290] This has proven to be a very successful solution, so much so that it was carried over into Advanced Virgo with little change.

Initial LIGO was built with a deliberately conservative design that used passive isolation stacks. Its performance was good for frequencies above 40 Hz, but not lower.[291] That served well enough for initial LIGO, but in Advanced LIGO a completely different system was implemented, with the aim of allowing good gravitational wave sensitivity down to 10 Hz.

The aLIGO system involves a quadruple pendulum for each of the test masses, with the upper stages including maraging steel blade springs for vertical isolation, as inspired by the Virgo design. A distinctive feature is that this suspension is mounted from an active vibration isolation system (Section 11.7.2) of breathtaking design. The active system is actually two platforms in series; each platform has inertial sensors for all six degrees of freedom. Horizontal degrees of freedom are tricky, because tilt and horizontal acceleration can't be distinguished by an inertial sensor. The solution was to combine measurements from inertial sensors on the platforms with predictions ("feedforward") based on additional sensors on the ground. Each active platform gives an order of magnitude of isolation in the band from 1 to

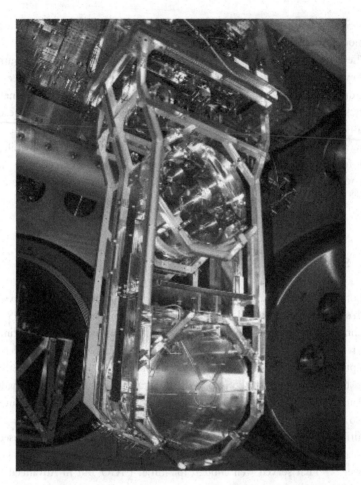

Figure 17.4: Advanced LIGO test mass suspension. (Figure Credit: Caltech/MIT/LIGO Lab)

10 Hz.[292] Performance below the bottom of the signal band at 10 Hz was a crucial design goal; in initial LIGO, large low-frequency motions of the mirrors dominated the required dynamic range of interferometer control systems, making it precarious to keep the interferometer locked and aligned properly at the operating point. The new aLIGO system has yielded excellent isolation performance and much more robust locking.

It should be noted that direct coupling of seismic noise is far from the only noise source at low frequencies; cross-coupling from many auxiliary degrees of freedom of the interferometers' complex optics and control systems also play an important role.

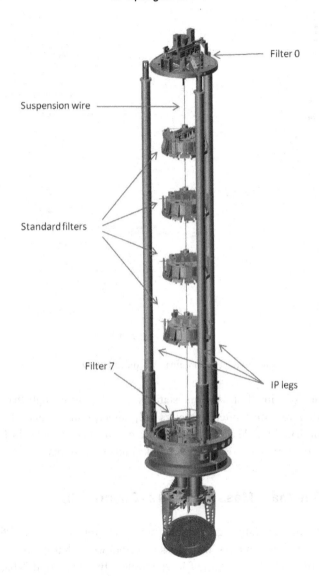

Filter 0

Suspension wire

Standard filters

Filter 7

IP legs

Figure 17.5: Advanced Virgo Super Attenuator. (Credit: Virgo)

Newtonian gravitational coupling of seismic noise to the test masses has been studied with much greater sophistication than in the references cited in Chapter 8.[293] In the aLIGO noise budget, Newtonian noise (now the preferred name) will be almost as important as seismic noise near 10 Hz, once noise from auxiliary degrees of freedom is sufficiently tamed.

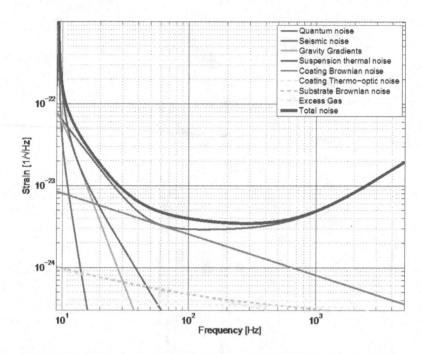

Figure 17.6: Noise budget for nominal Advanced LIGO design. (Credit: LIGO)

Seismic noise in all of its manifestations is important enough that it may be necessary to move future interferometers deep underground, where the noise can be orders of magnitude lower.[294] The new KAGRA interferometer in Japan, now under construction, is located in the Kamioka mine for just this reason.

17.10 Resonant Mass Detectors (Chapter 13)

In 1994, the best operating gravitational wave detectors weren't interferometers at all; they were cryogenic resonant mass detectors whose lineage could be traced directly to the original detectors of Joseph Weber. But by around 2003, the early performance of the new km-scale interferometers had surpassed the sensitivity of the best resonant detectors. Now, the resonant mass detectors are no longer observing. The transition was marked by a joint analysis of data from the LIGO Livingston interferometer and the ALLEGRO bar at nearby Louisiana State University, in a search for a ~900 Hz stochastic signal.[295]

A great deal of cleverness is evident in these detectors; there are both physics and engineering lessons to learn from them. They repay careful study even today.

Figure 17.7: Aerial view of LIGO Livingston Observatory. (Credit: Caltech/MIT/LIGO Lab)

17.11 Large Interferometers (Chapters 9 and 16)

Nothing exemplifies the changes since 1994 more dramatically than the topics treated in Chapters 9 and 16. Expressed there as a set of "design features" and "prospects", we are now able to look back on the successful realization of the first generation of interferometers and to study the second generation versions which have achieved scientific success even before fully reaching their design performance.

The length scale of 3 or 4 km proved to allow sufficient sensitivity (not dominated by seismic or thermal displacements of the mirrors) at frequencies of 30 Hz and higher (so far), which by good fortune allowed the discovery of signals that swept up through this band into the hundreds of hertz. The large capital expense for remote observatories with long vacuum systems at pressures below 10^{-8} torr was worth it — they worked!

Newcomers should realize that these instruments are substantially more complex than could be explained in a beginner's book such as this. The design features are summarized in Reference 278. While the interferometer sketched in, say, Figure 12.10 involved seven crucial suspended optics (two mirrors for each of the two long Fabry-Perot cavities, the beamsplitter, the power recycling mirror and the

Figure 17.8:   Aerial view of Virgo. (Credit: European Gravitational Observatory)

signal recycling mirror), a working interferometer such as aLIGO has about two dozen suspended optics. There are hundreds of feedback loops (most of considerable complexity and most of multi-input-multi-output form). Another measure of complexity is the number of degrees of freedom that are measured and recorded: for each aLIGO interferometer there are over 200 channels recorded at a bandwidth of 16 kilosamples/sec or greater, and over 200,000 channels archived overall. Many of these channels record signals from the set of environmental monitors that are crucial to demonstrating that a signal was not caused by a local disturbance.

There are already glimpses of what the next generation of interferometers will look like. The Japanese KAGRA interferometer is pioneering two potentially essential features: underground installation and cryogenic temperatures to reduce thermal noise.[296] The most complete design study of a possible third generation interferometer is that of the Einstein Telescope.[297] Like KAGRA, it is envisioned to be underground and cryogenic; it also will have 10-km long arms in a triangular configuration for measuring both polarizations of a gravitational wave. Other visions are also being developed.[298] We live in interesting times!

17.12   Data Analysis (Chapter 14)

A newcomer will be well served by study of the basic ideas of data analysis described in Chapter 14, but she will soon find the need for more. Sophisticated methods have now been developed for four major classes of signals:

- compact binary inspirals that, because they can be accurately modeled, are sought using matched filtering,[299]
- other transients that can't rely on templates and that are found by searching for brief excursions of band-limited signal power,[300]
- periodic signals from rotating neutron stars, for which the basic search tool is the Fourier transform (modified to account for the various modulations of the otherwise perfectly sinusoidal signal),[301] and

- a stochastic background, found via the cross-correlation of multiple spatially-separated interferometers.[302]

The textbook by Creighton and Anderson does a much better job than the present book in discussing data analysis. A reader who wants to understand the essential ideas at a deeper level will be best served by diving into some of the recent literature in the papers cited above, and following the chains of references.

Not surprisingly, the noise from gravitational wave interferometers is not perfectly Gaussian, but contains (not infrequent enough) glitches that complicate data analysis. Two lines of attack have proven successful in minimizing the impact of this problem on the sensitivity of searches. For compact binaries with well-predicted signals, each trigger is checked for its fidelity to the signal model, using a chi-squared test.[303] In addition, a robust set of data quality investigations find and flag times of corrupted data output (for the transient searches) and mark the frequencies that should be ignored in frequency domain searches.[304, 305]

17.13 Gravitational Wave Astronomy (Chapter 15)

Now that the very first gravitational wave signals have been found, we have entered the era of gravitational wave astronomy. (How gratifying it is to be able to write that sentence!) By a fluke of history, the first signals were found by the two LIGO interferometers alone, but by the time you read these words we firmly expect the international network to include Advanced Virgo. Within a decade or sooner, KAGRA and LIGO-India will fill out the network.

The building up of the global network is essential for precisely the reasons explained in Section 15.1 — to allow the localization of the source position to a small region on the sky. The first detected signal, GW150914, had a position error box whose area was 620 square degrees (at 90% confidence)! Even though only the two LIGO interferometers were available to detect the signal, the position was constrained to only a portion of the 7 msec time delay's ring on the sky, by use of the ratio of the signals detected at the two sites.[306] But a full global network will do much better; four well-placed sites around the world could yield error boxes on the sky with typical sizes of a few to a few tens of square degrees.[307]

One of the most important reasons to look forward to tighter position error boxes is that it will aid in the identification of electromagnetic signals that may also be detected from the sources that generated the gravitational waves. While black hole binaries in a pristine environment ought to generate gravitational wave signals but nothing else, systems involving neutron stars (for example) will produce very informative electromagnetic signals.

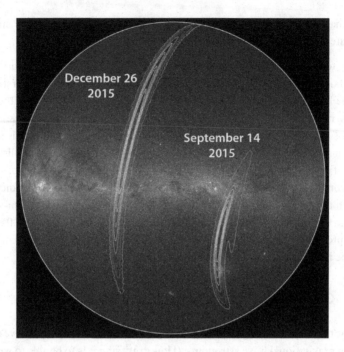

Figure 17.9: Sky position error boxes for GW150914 and for GW151226. (Credit: LIGO/Axel Mellinger)

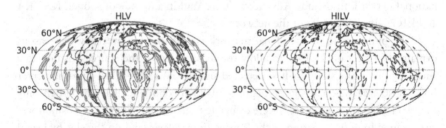

Figure 17.10: Position error boxes on the sky for a LIGO/Virgo network and for a network that also adds LIGO-India. For more information, see Reference 307.

Already in the initial interferometer era, LIGO and Virgo developed the techniques to receive alerts of gamma ray bursts (as well as of other electromagnetic transients), and to make intensive searches for coincident gravitational wave signals.[308] By the start of the advanced detector era, arrangements were made for gravitational wave detectors to generate alerts to electromagnetic astronomers. Follow-up of huge error boxes from black-hole-only sources was not expected to yield a coincidence, but it provided an excellent shake-down of procedures and taught us lessons about how best to share information. For an account of these first efforts, see Reference 306.

Of course, gravitational wave observers can do unique science using only the gravitational wave signal itself. The key reason is given in Eq. (3.6). Unlike most sources of electromagnetic radiation in the Universe, gravitational waves like GW150914 are generated by a single coherent system rather than by an incoherent superposition of many independent elementary radiators. Thus, the time series of $h(t)$ can be interpreted directly (except in the most dramatically relativistic regime) as the time series of the second time derivative of the mass quadrupole moment of the source. For many kinds of sources, the signal will invite a single unique interpretation. GW150914 was precisely such a case; even if we hadn't been searching for it with a set of matched templates, the time series of the signal would have strongly suggested that we were observing the last few orbits of a binary (and told us about the masses involved) and the first few cycles of vibration of the resultant combined object.[300]

Fortunately, we did have very well developed models to compare against GW 150914. That enabled a fairly precise determination of the masses of the initial black holes and of the final one,[309] as well as the first tests of the physics of general relativity in the strong field regime.[310] In addition, the first-ever observations of a black hole binary give unique new input to the astronomy of star formation and the evolution of binaries.[311]

As with any snapshot of a scientific field in its early stages, it will not be long before these paragraphs seem as quaint as some of the sections of the first edition of this book. We also look forward to the time when the field of gravitational wave detection is enriched by discoveries from space-based interferometers,[312] pulsar timing arrays,[313] and measurements of the polarization of the cosmic microwave background.[314]

May those days come soon!

References

1. R. H. Dicke, *The Theoretical Significance of Experimental Relativity* (Gordon and Breach, New York, 1964).
2. See the excellent Einstein biography by A. Pais, *'Subtle is the Lord...'* (Clarendon Press, Oxford, 1982).
3. J. Weber, *Phys. Rev.* **117** (1960) 306.
4. J. Weber, *Phys. Rev. Lett.* **22** (1969) 1320.
5. J. A. Tyson and R. P. Giffard, *Ann. Rev. Astron. Astrophys.* **16** (1978) 521.
6. A. Abramovici, W. E. Althouse, R. W. P. Drever, Y. Giirsel, S. Kawamura, F. J. Raab, D. Shoemaker, L. Sievers, R. E. Spero, K. S. Thorne, R. E. Vogt, R. Weiss, S. E. Whitcomb, and M. E. Zucker, *Science* **256** (1992) 325.
7. C. Bradaschia, R. Del Fabbro, A. DiVirgilio, A. Giazotto, H. Kautzky, V. Montelatici, D. Passuello, A. Brillet, O. Cregut, P. Hello, C. N. Man, P. T. Manh, A. Marraud, D. Shoemaker, J.-Y. Vinet, F. Barone, L. DiFiore, L. Milano, G. Russo, S. Solimeno, J. M. Aguirregabiria, H. Bel, J.-P. Duruisseau, G. Le Denmat, P. Tourrenc, M. Capozzi, M. Longo, M. Lops, I. Pinto, G. Rotoli, T. Damour, S. Bonazzola, J. A. Marek, Y. Gourghoulon, L. E. Holloway, F. Fuligni, V. Iafolla, G. Natale, *Nucl. Instr. Meth. Phys. Res.* **A289** (1990) 518.
8. M. Harwit, *Cosmic Discovery* (Basic Books, New York, 1981).
9. K. S. Thorne, in *300 Years of Gravitation*, eds. S.W. Hawking and W. Israel (Cambridge University Press, Cambridge, 1987).
10. D. G. Blair, ed., *The Detection of Gravitational Waves* (Cambridge University Press, Cambridge, 1991).
11. See, e.g., C. W. Misner, K. S. Thorne, and J. A. Wheeler, *Gravitation*, (W. H. Freeman, San Francisco, 1973).
12. E. F. Taylor and J. A. Wheeler, *Spacetime Physics* (Freeman, San Francisco, 1966).
13. A. A. Michelson and E. W. Morley, *Am. J. Sci.* **34** (1887) 333. This paper, and two previous ones, are included as appendices in the valuable history by Loyd S. Swenson, *The Ethereal Aether* (University of Texas, Austin, 1972).
14. H. A. Haus, *Waves and Fields in Optoelectronics* (Prentice-Hall, Englewood Cliffs, New Jersey, 1984).
15. M. E. Gertsenshtein and V. I. Pustovoit, *Sov. Phys. JETP* **16** (1962) 433.

16. G. E. Moss, L. R. Miller, and R. L. Forward, *Appl. Opt.* **10** (1971) 2495; R. L. Forward, *Phys. Rev.* **D17** (1978) 379.
17. R. Weiss, *Quarterly Progress Report, MIT Research Lab of Electronics* **105** (1972) 54.
18. B. F. Schutz and M. Tinto, *M.N.R.A.S.* **224** (1987) 131.
19. R. Spero, "Radiation Sensitivity Pattern", unpublished report, 1983.
20. S. P. Boughn, private communication.
21. P. Kafka, in *'Space Science and Fundamental Physics'*, *Proceedings of Summer School held at Alpbach, Austria, 28 July–7 August 1987* (ESA SP-283, May 1988).
22. M. S. Longair, *Theoretical Concepts in Physics* (Cambridge University Press, Cambridge, 1984) pp. 37–52.
23. J. H. Bryant, *Heinrich Hertz: The Beginning of Microwaves* (IEEE, New York 1988).
24. D. Cuomo, G. Franceschetti, G. Panariello, I. M. Pinto, S. P. Petracca, S. Amata, and G. Rotoli, in *Experimental Gravitational Physics*, ed. P. F. Michelson (World Scientific, Singapore, 1988).
25. E. Teller, talk to NATO ASI on *Cosmology and Particle Physics*, Erice, 1993.
26. R. A. Hulse and J. H. Taylor, *Ap. J. Lett.* **195** (1975) L51.
27. J. H. Taylor and J. M. Weisberg, *Ap. J.* **253** (1982) 908; J. H. Taylor and J. M. Weisberg, *Ap. J.* **345** (1989) 434.
28. A. H. Batten, *Binary and Multiple Systems of Stars* (Pergamon, Oxford, 1973), pp. 100 ff.
29. D. H. Douglass and V. B. Braginsky, in *General Relativity: an Einstein Centenary Survey*, eds. S.W. Hawking and W. Israel (Cambridge University Press, Cambridge, 1979).
30. P. C. Peters and J. Mathews, *Phys. Rev.* **131** (1963) 435.
31. The history of Pauli's proposal is well told in A. Pais, *Inward Bound* (Clarendon Press, Oxford, 1986).
32. C. D. Ellis, *Proc. Roy. Soc.* **A161** (1937) 447, cited in A. Pais, *ibid.*
33. F. Reines, *Ann. Rev. Nucl. Sci.* **10** (1960) 1.
34. J. P. A. Clark and D. M. Eardley, *Ap. J.* **215** (1977) 315.
35. J. P. A. Clark, E. P. J. van den Heuvel, and W. Sutantyo, *Astron. Astrophys.* **72** (1979) 120.
36. E. S. Phinney, *Ap. J. Lett.* **380** (1991) L17; see also R. Narayan, T. Piran, and A. Shemi, *Ap. J. Lett.* **379** (1991) L17.
37. V. Trimble, *Rev. Mod. Phys.* **54** (1982) 1183.
38. W. D. Arnett, J. N. Bahcall, R. P. Kirshner, and S. E. Woosley, *Ann. Rev. Astron. Astrophys.* **27** (1989) 629.
39. S. E. Woosley and T. A. Weaver, *Ann. Rev. Astron. Astrophys.* **24** (1986) 205.
40. K. Hirata *et al.*, *Phys. Rev. Lett.* **58** (1987) 1490; R. M. Bionta *et al.*, *Phys. Rev. Lett.* **58** (1987) 1494.
41. R. A. Saenz and S. L. Shapiro, *Ap. J.* **244** (1981) 1032.
42. A. Hewish, S. J. Bell, J. D. H. Pilkington, P. F. Scott, and R. A. Collins, *Nature* **217** (1968) 709; and *Nature* **218** (1968) 126.
43. T. Gold, *Nature* **218** (1968) 731.
44. F. G. Smith, *Pulsars* (Cambridge University Press, Cambridge, 1977); see also R. N. Manchester and J. H. Taylor, *Pulsars* (W. H. Freeman, San Francisco, 1977).
45. J. P. Ostriker and J. E. Gunn, *Ap. J.* **157** (1969) 1395.
46. M. A. Alpar, A. F. Cheng, M. A. Ruderman, and J. Shaham, *Nature* **300** (1982) 728.
47. R. V. Wagoner, *Ap. J.* **278** (1984) 345.
48. S. Chandrasekhar, *Phys. Rev. Lett.* **24** (1970) 611.

49. J. L. Friedman and B. F. Schutz, *Ap. J.* **222** (1978) 281.
50. L. Lindblom, *Ap. J.* **303** (1986) 146; C. Cutler and L. Lindblom, *Ap. J.* **314** (1987) 234.
51. W. Israel, in *300 Years of Gravitation*, eds. S.W. Hawking and W. Israel (Cambridge University Press, Cambridge, 1987).
52. A. P. Cowley, *Ann. Rev. Astron. Astrophys.* **30** (1992) 287.
53. P. J. Young, J. A. Westphal, J. Kristian, C. P. Wilson, and F. P. Landauer, *Ap. J.* **221** (1978) 721.
54. R. Genzel and C. H. Townes, *Ann. Rev. Astron. Astrophys.* **25** (1987) 377.
55. R. D. Blandford, in *Sources of Gravitational Radiation*, ed. L. Smarr (Cambridge University Press, Cambridge, 1979).
56. M. J. Rees, in *Structure and Properties of Nearby Galaxies*, ed. Berkhuijsen and Wiekbinskii (D. Reidel, Dordrecht, 1978).
57. S. L. Detweiler and E. Szedenits, Jr., *Ap. J.* **231** (1979) 211.
58. L. Smarr, in *Sources of Gravitational Radiation*, ed. L. Smarr (Cambridge University Press, Cambridge, 1979).
59. B. J. Carr, *Astron. Astrophys.* **89** (1980) 6.
60. L. P. Grishchuk, *Annals of NY Acad. Sci.* **302** (1977) 439; A. A. Starobinsky, *JETP Lett.* **30** (1979) 682.
61. W. B. Davenport, Jr., and W. L. Root, *An Introduction to the Theory of Random Signals and Noise* (McGraw-Hill, New York, 1958).
62. R. N. Bracewell, *The Fourier Transform and Its Applications* (McGraw-Hill, New York, 1986).
63. E. O. Brigham, *The Fast Fourier Transform* (Prentice-Hall, Englewood Cliffs, New Jersey, 1974).
64. J. W. Cooley and J. W. Tukey, *Math. Comput.* **19** (1965) 297; E. O. Brigham, *ibid.*
65. A. B. Pippard, *Response and Stability* (Cambridge University Press, Cambridge, 1985).
66. G. F. Franklin, J. D. Powell, and A. Emami-Naeini, *Feedback Control of Dynamic Systems* (Addison-Wesley, Reading, Massachusetts, 1987).
67. W. B. Davenport, Jr. and W. L. Root, *ibid.*, pp. 154–7.
68. W. B. Davenport, Jr. and W. L. Root, *ibid.*, Chapter 11.
69. L. A. Wainstein and V. D. Zubakov, *Extraction of Signals from Noise* (Dover, New York, 1970), Chapter 3.
70. A. D. Whalen, *Detection of Signals in Noise* (Academic, New York, 1971), Chapter 6.
71. W. B. Davenport, Jr. and W. L. Root, *ibid.*, Chapter 7.
72. D. Shoemaker, A. Brillet, C. N. Man, O. Crégut, and G. Kerr, *Optics Letters* **14** (1989) 609.
73. N. Bohr, *Nature* **121** (1928) 580; reprinted in J. A. Wheeler and W. H. Zurek, eds., *Quantum Theory and Measurement* (Princeton University Press, Princeton, New Jersey, 1983).
74. C. M. Caves, in *Quantum Measurement and Chaos*, eds. E. R. Pike and S. Sarkar (Plenum, New York, 1987).
75. P. A. M. Dirac, *The Principles of Quantum Mechanics* (Clarendon Press, Oxford, 1958).
76. R. Weiss, in *Sources of Gravitational Radiation*, ed. L. Smarr (Cambridge University Press, Cambridge, 1979) p. 7.
77. C. M. Caves, *Phys. Rev. Lett.* **45** (1980) 75.

78. A. Brillet, J. Gea-Banacloche, G. Leuchs, C. N. Man and J. Y. Vinet, in *The Detection of Gravitational Waves*, ed. D. G. Blair (Cambridge University Press, Cambridge, 1991).

79. D. Shoemaker, R. Schilling, L. Schnupp, W. Winkler, K. Maischberger, and A. Rüdiger, *Phys. Rev.* **D38** (1988) 423.

80. See the description in A. Yariv, *Optical Electronics* (Holt, Rinehart and Winston, New York 1985).

81. D. Herriott, H. Kogelnik, and R. Kompfner, *Appl. Opt.* **3** (1964) 523.

82. P. S. Linsay, in R. Weiss, P. S. Linsay, and P. R. Saulson, *A Study of a Long Baseline Gravitational Wave Antenna System* (MIT, Cambridge, Mass., 1983), unpublished.

83. H. A. Haus, *ibid.*, pp. 108 ff.

84. H. A. Haus, *ibid.*, pp. 122 ff.

85. A. Rüdiger, R. Schilling, L. Schnupp, W. Winkler, H. Billing, and K. Maischberger, *Optica Acta* **28** (1981) 641.

86. R. Schilling, L. Schnupp, W. Winkler, H. Billing, K. Maischberger, and A. Rüdiger, *J. Phys. E: Sci. Instrum.* **14** (1981) 65.

87. L. Schnupp, W. Winkler, K. Maischberger, A. Rudiger, and R. Schilling, *J. Phys. E: Sci. Instrum.* **18** (1985) 482.

88. R. W. P. Drever, G. M. Ford, J. Hough, I. M. Kerr, A. J. Munley, J. R. Pugh, N. A. Robertson, and H. Ward, in *Proceedings of the Ninth International Conference on General Relativity and Gravitation (Jena 1980)*, ed. E. Schmutzer (VEB Deutscher Verlag der Wissenschaften, Berlin, 1983).

89. K. S. Thorne, California Institute of Technology preprint GRP-200, 1989.

90. R. Weiss, P. S. Linsay, and P. R. Saulson, *A Study of a Long Baseline Gravitational Wave Antenna System* (1983) unpublished.

91. See, e.g., B. J. Meers, *Phys. Rev.* **D38** (1988) 2317.

92. R. Brown, *Phil. Mag.* **4** (1828) 161; *Ann. Phys. Chern.* **14** (1828) 294.

93. D. K. C. MacDonald, *Noise and Fluctuations: An Introduction* (Wiley, New York, 1962).

94. J. Perrin, *Atoms* (Ox Bow Press, Woodbridge, Connecticut, 1990).

95. A. Einstein, *Investigations on the Theory of the Brownian Movement* (Dover, New York, 1956).

96. F. Reif, *Fundamentals of Statistical and Thermal Physics* (McGraw-Hill, New York, 1965).

97. V. B. Braginsky, V. P. Mitrofanov, and V.I. Panov, *Systems with Small Dissipation* (University of Chicago Press, Chicago, 1977).

98. H. B. Callen and T. A. Welton, *Phys. Rev.* **83** (1951) 34; H. B. Callen and R. F. Greene, *Phys. Rev.* **86** (1952) 702.

99. J. B. Johnson, *Phys. Rev.* **32** (1928) 97.

100. H. Nyquist, *Phys. Rev.* **32** (1928) 110.

101. The crucial definite integral is number 3.242 in I. S. Gradshteyn and I. M. Ryzhik, *Table of Integrals, Series, and Products* (Academic, New York, 1980).

102. E. M. Purcell, *Electricity and Magnetism* (McGraw-Hill, New York, 1965) pp. 276, 284.

103. C. Zener, *Elasticity and Anelasticity of Metals* (University of Chicago Press, Chicago, 1948).

104. A. L. Kimball and D. E. Lovell, *Phys. Rev.* **30** (1927) 948; see also Braginsky, Mitrofanov, and Panov, *ibid.*

105. A. S. Nowick and B.S. Berry, *Anelastic Relaxations in Crystalline Solids* (Academic, New York, 1972).
106. A. L. Kimball and D. E. Lovell, *ibid*.
107. F. K. du Pre, *Phys. Rev.* **78** (1950) 615.
108. J. L. Routbort and H. S. Sack, *J. Appl. Phys.* **37** (1966) 4803.
109. P. R. Saulson, *Phys. Rev.* **D42** (1990) 2437.
110. L. D. Landau and E. M. Lifshitz, *Theory of Elasticity* (Pergamon, Oxford, 1970) pp. 89 ff.
111. P. R. Saulson, *ibid.;* A. Gillespie and F. Raab, *Phys. Lett.* **A178** (1993) 357.
112. R. Weiss, unpublished, cited in P.R. Saulson, *ibid.;* G. I. GonzaJez and P. R. Saulson, *J. Acoust. Soc. Am.* (1994); J. E. Logan, J. Hough, and N. A. Robertson, *Phys. Lett. A* (1993).
113. G. W. MacMahon, *J. Acoust. Soc. Am.* **36** (1964) 85.
114. J. R. Hutchinson, *J. Appl. Mech.* **47** (1980) 901.
115. Mitrofanov and Frontov (1974), cited by Braginsky, Mitrofanov and Panov, *ibid*.
116. A. Gillespie and F. Raab, paper in preparation, 1993.
117. M. Bath, *Introduction to Seismology* (Birkhauser, Basel, 1979).
118. T. L. Aldcroft, P. F. Michelson, R. C. Taber, and F. A. McLoughlin, *Rev. Sci. Instrum.* **63** (1992) 3815.
119. N. A. Robertson, in *The Detection of Gravitational Waves*, ed. D. G. Blair (Cambridge University Press, Cambridge, 1991).
120. A stack, without detailed specifications, appears in the diagram of the apparatus in J. Weber, *Phys. Rev. Lett.* **17** (1966) 1228.
121. See, e. g., Nowick and Berry, *ibid.*, pp. 37–8.
122. J. Giaime, P. Saha, D. Shoemaker, and L. Sievers, paper in preparation, 1993.
123. R. Del Fabbro, A. diVirgilio, A. Giazotto, H. Kautzky, V. Montelatici, and D. Passuello, *Phys. Lett.* **A132** (1988) 237.
124. A. Giazotto, private communication.
125. R. E. Spero, in *Science Underground*, ed. M. M. Nieto *et al.* (AlP, New York, 1983).
126. P. R. Saulson, *Phys. Rev.* **D30** (1984) 732.
127. R. T. Stebbins, talk at 12th International Conference on General Relativity and Gravitation, Boulder, CO, 1989.
128. A. Abramovici, private communication, 1992.
129. R. Weiss, in R. Weiss, P. S. Linsay, and P. R. Saulson, *A Study of a Long Baseline Gravitational Wave Antenna System* (1983) unpublished.
130. A. Rudiger, unpublished report, 1981; S. Whitcomb, unpublished report, 1984.
131. K. Danzmann, A. Rudiger, R. Schilling, W. Winkler, J. Hough, G. P. Newton, D. Robertson, N. A. Robertson, H. Ward, P. Bender, D. Hils, R. Stebbins, C. D. Edwards, W. Falkner, M. Vincent, A. Bernard, B. Bertotti, A. Brillet, C. N. Man, M. Cruise, P. Gray, M. Sandford, R. W. P. Drever, V. Kose, M. Kuhne, B. F. Schutz, R. Weiss, and H. Welling, *LISA: Proposal for a Laser-Interferometric Gravitational Wave Detector in Space*, (Max-Planck-Institut fiir Quantenoptik, Garching, 1993) Report 177, unpublished.
132. B. Lange, *AIAA Journal* **2** (1964) 1590.
133. P. G. Roll, R. Krotkov, and R. H. Dicke, *Ann. Phys. (N. Y.)* **26** (1964) 442.

134. V. B. Braginsky and V.I. Panov, *Sov. Phys. JETP* **34** (1972) 463; V. B. Braginsky and A. B. Manukin, *Measurement of Weak Forces in Physics Experiments* (University of Chicago Press, Chicago, 1977).

135. See, e. g., J. Hough, H. Ward, G. A. Kerr, N. L. MacKenzie, B. J. Meers, G. P. Newton, D. I. Robertson, N. A. Robertson, and R. Schilling, in *The Detection of Gravitational Waves*, ed. D. G. Blair (Cambridge University Press, Cambridge, 1991).

136. R. H. Dicke, *Rev. Sci. Instrum.* **17** (1946) 268.

137. J. C. Mather, E. S. Cheng, R. E. Eplee, Jr., R. B. Isaacman, S. S. Meyer, R. A. Shafer, R. Weiss, E. L. Wright, C. L. Bennett, N. W. Boggess, E. Dwek, S. Gulkis, M.G. Hauser, M. Janssen, T. Kelsall, P.M. Lubin, S. H. Moseley, Jr., T. L. Murdock, R.F. Silverberg, G. F. Smoot, and D. T. Wilkinson, *Ap. J. Lett.* **354** (1990) L37.

138. R. J. Bell, *Introductory Fourier Transform Spectroscopy* (Academic, New York, 1972).

139. W. H. Press, *Comments Astrophys.* **7** (1978) 103.

140. M. B. Weissman, *Rev. Mod. Phys.* **60** (1988) 537.

141. W. H. Press, *Gen. Rel. Grav.* **11** (1979) 105.

142. W. Winkler, in *The Detection of Gravitational Waves*, ed. D. G. Blair (Cambridge University Press, Cambridge, 1991).

143. O. Mayr, *Origins of Feedback Control* (MIT Press, Cambridge, 1970).

144. See, e. g., G. B. Thomas, Jr., *Elements of Calculus and Analytic Geometry* (Addison-Wesley, Reading, Mass., 1967).

145. C. T. Molloy, *J.Acoust. Soc. Am.* **29** (1957) 842.

146. M. E. Zucker, private communication, cifed in J. Hough *et al.*, in *The Detection of Gravitational Waves*, ed. D. G. Blair (Cambridge University Press, Cambridge, 1991).

147. See, e.g., A. B. Pippard, *The Physics of Vibration* (Cambridge University Press, Cambridge, 1989), pp. 107 ff.

148. G. F. Franklin, J.D. Powell, and A. Emami-Naeni, *ibid.*, pp. 258 ff.

149. I. M. Horowitz, *Synthesis of Feedback Systems* (Academic, New York, 1963).

150. P. Horowitz and W. Hill, *The Art of Electronics* (Cambridge University Press, Cambridge, 1989) pp. 180 ff.

151. H. Heffner, *Proc. IRE* **50** (1962) 1604.

152. R. W. P. Drever, in *The Detection of Gravitational Waves*, ed. D. G. Blair (Cambridge University Press, Cambridge, 1991).

153. P. R. Saulson, in *Gravitational Astronomy: Instrument Designs and Astrophysical Prospects*, eds. D. E. McClelland and H.-A. Bachor (World Scientific, Singapore, 1991), and references therein.

154. D. B. Newell, P. L. Bender, J. E. Faller, P. G. Nelson, S. J. Richman, and R. T. Stebbins, talk at 13th International Conference on General Relativity and Gravitation, Huerta Grande, Cordoba, Argentina, 1992.

155. H. A. Haus, *ibid.*, pp. 327 ff.

156. *Reference Data for Radio Engineers* (Howard W. Sams, Indianapolis, Indiana, 1975).

157. D. C. Green, *Radio Systems Technology* (Longman Scientific & Technical, Harlow, Essex, 1990).

158. T. M. Niebauer, R. Schilling, K. Danzmann, A. Rüdiger, and W. Winkler, *Phys. Rev.* **A43** (1990) 5022.

159. D. Dewey, Ph. D. thesis (MIT, Cambridge, Mass., 1986).

160. R. W. P. Drever, J. L. Hall, F. V. Kowalski, J. Hough, G. M. Ford, A. J. Munley, and H. Ward, *Appl. Phys.* **B31** (1983) 97; A. Schenzle, R. DeVoe, and G. Brewer, *Phys. Rev.* **A25** (1982) 2606.

161. R. V. Pound, *Rev. Sci. Instrum.* **17** (1946) 490.

162. R. W. P. Drever, in *The Detection of Gravitational Waves, ibid.*

163. R. W. P. Drever, in *Gravitational Radiation*, eds. N. Deruelle and T. Piran (North Holland, Amsterdam, 1983) p. 321.

164. D. Shoemaker, P. Fritsche!, J. Giaime, N. Christensen, and R. Weiss, *Appl. Opt.* **30** (1991) 3133.

165. B. J. Meers, *Phys. Rev.* **D38** (1988) 2317; K. A. Strain and B. J. Meers, *Phys. Rev. Lett.* **66** (1991) 1391.

166. J. Mizuno, K. A. Strain, P. G. Nelson, J. M. Chen, R. Schilling, A Rüdiger, W. Winkler, and K. Danzmann, *Phys. Lett.* **A175** (1993) 273.

167. P. F. Michelson, J. C. Price, and R. C. Taber, *Science* **237** (1987) 150.

168. C. D. B. Bryan, *The National Air and Space Museum* (H. N. Abrams, New York, 1988).

169. L. H. Sullivan, *Lippincott's Magazine* **57** (1896) 403. This remarkable essay, "The Tall Office Building Artistically Reconsidered," is reprinted in *Louis Sullivan: The Public Papers*, ed. R. Twombly (University of Chicago Press, Chicago, 1988).

170. H. J. Paik, *J. Appl. Phys.* **41** (1976) 1168.

171. L. Meirovitch, *Elements of Vibration Analysis* (McGraw-Hill, New York, 1975).

172. J.-P. Richard, in *The Detection of Gravitational Waves*, ed. D. G. Blair (Cambridge University Press, Cambridge, 1991).

173. J.-P. Richard, *Phys. Rev. Lett.* **52** (1984) 165.

174. J. C. Price, *Phys. Rev.* **D36** (1987) 3555.

175. G. W. Gibbons and S. W. Hawking, *Phys. Rev.* **D4** (1971) 2191.

176. J. L. Levine and R. L. Garwin, *Phys. Rev. Lett.* **31** (1973) 173.

177. E. Amaldi, O. Aguiar, M. Bassan, P. Bonifazi, P. Carelli, M.G. Castellano, G. Cavallari, E. Coccia, C. Cosmelli, W. M. Fairbank, S. Frasca,V. Foglietti, R. Habel, W. O. Hamilton, J. Henderson, W. Johnson, K. R. Lane, A. G. Mann, M.S. McAshan, P. F. Michelson, I Modena, G. V. Pallottino, G. Pizzella, J. C. Price, R. Rapagnani, F. Ricci, N. Solomonson, T.R. Stevenson, R. C. Taber, and B.-X. Xu, *Astron. Astrophys.* **216** (1989) 325.

178. N. Solomonson, O. Aguiar, Z. Geng, W. O. Hamilton, W. W. Johnson, S. Merkowitz, B. Price, B. Xu, and N. Zhu, in *The Sixth Marcel Grossman Meeting*, eds. H. Sato and T. Nakamura (World Scientific, Singapore, 1992).

179. J.-P. Richard, *Phys. Rev.* **D46** (1992) 2309.

180. H. J. Paik, *ibid.;* see the review by J.-P. Richard in *The Detection of Gravitational Waves*, ed. D. G. Blair (Cambridge University Press, Cambridge, 1991).

181. H. Rothe and W. Dahlke, *Proc. IRE* **44** (1956) 811.

182. F. V. Hunt, *Electroacoustics* (Harvard University Press, Cambridge, Mass., 1954).

183. R. P. Giffard, *Phys. Rev.* **D14** (1976) 2478.

184. E.-K. Hu, C. Zhou, L. Mann, P. F. Michelson, and J. C. Price, *Phys. Lett.* **A157** (1991) 209.

185. W. W. Johnson and S.M. Merkowitz, *Phys. Rev. Lett.* **70** (1993) 2367.

186. See V. B. Braginsky and A. B. Manukin, *Measurement of Weak Forces in Physics Experiments* (University of Chicago Press, Chicago, 1977).

187. C. M. Caves, K. S. Thorne, R. W. P. Drever, V. D. Sandberg, and M. Zimmerman, *Rev. Mod. Phys.* **52** (1980) 341.

188. M. F. Bocko and W. W. Johnson, *Phys. Rev.* **A30** (1984) 2135.

189. R. E. Slusher, L. W. Hollberg, B. Yurke, J. C. Mertz, and J. F.Valley, *Phys. Rev. Lett.* **55** (1985) 2409.

190. O. Dahlman and H. Israelson, *Monitoring Underground Nuclear Explosions* (Elsevier, Amsterdam, 1977).

191. J.P. Ostriker, private communication.

192. F. Reif, *ibid.*, pp. 37 ff.

193. See, e. g., M. Abramowitz and I. A. Stegun, *Handbook of Mathematical Functions* (N. B. S., Washington, 1964); see also G. L. Squires, *Practical Physics* (Cambridge University Press, Cambridge, 1985) pp. 2,3–5 and p. 176.

194. B. F. Schutz, private communication, 1987; see B. F. Schutz and M. Tinto, *ibid.*

195. R. W. P. Drever, J. Hough, A. J. Munley, S.-A. Lee, R. Spero, S. E. Whitcomb, J. Pugh, G. Newton, B. J. Meers, E. Brooks III, and Y. Gürsel, in *Quantum Optics, Experimental Gravity, and Measurement Theory*, eds. S. Meystere and M. O. Scully (Plenum, New York, 1983) p. 503.

196. J. C. Livas, Ph. D. Thesis (MIT, Cambridge Mass., 1987).

197. B. F. Schutz, in *The Detection of Gravitational Waves*, ed. D. G. Blair (Cambridge University Press, Cambridge, 1991).

198. See, e. g., R.N. Bracewell, *ibid.*, pp. 55–6.

199. See, e. g., D. C. Green, *ibid.*, pp. 24 ff.

200. See, e. g., the review by J. H. Taylor and D. R. Stinebring, *Ann. Rev. Astron. Astrophys.* **24** (1986) 285.

201. See the reminiscence by R. W. Wilson, *Physica Scripta* **21** (1980) 599.

202. N. Christensen, *Phys. Rev.* D46 (1992) 5250; N. Christensen, Ph. D. Thesis (MIT, Cambridge, Mass. 1990).

203. Y. Gürsel and M. Tinto, *Phys. Rev.* **D40** (1989) 3884.

204. R. Giacconi, in *X-Ray Astronomy*, eds. R. Giacconi and H. Gursky (D. Reidel, Dordrecht, 1974) pp. 155 ff.

205. See, e. g., D. H. Menzel and J. M. Pasachoff, *A Field Guide to the Stars and Planets* (Houghton Mifflin, Boston, 1990).

206. R. W. Klebesadel, I. B. Strong, and R. A. Olson, *Ap. J. Lett.* **182** (1973) L85; for a recent (but pre-Compton) review, see J. C. Higdon and R. E. Lingenfelter, *Ann. Rev. Astron. Astrophys.* **28** (1990) 401.

207. C. A. Meegan, G. J. Fishman, R. B. Wilson, W. S. Paciesas, G. N. Pendleton, J. M. Horack, M. N. Brock, and C. Kouveliotou, *Nature* **355** (1992) 143.

208. D. Eichler, M. Livio, T. Piran, and D. N. Schramm, *Nature* **340** (1989) 126.

209. C. S. Kochanek and T. Piran, *Ap. J. Lett.* **417** (1993) 117.

210. J. S. Hey, *The Evolution of Radio Astronomy* (Science History Publications, New York, 1973).

211. W. Tucker and R. Giacconi, *The X-Ray Universe* (Harvard University Press, Cambridge, Mass., 1985); R. Giacconi and H. Gursky, eds., *X-Ray Astronomy* (D. Reidel, Dordrecht, 1974).

212. N. Kawashima, in *Experimental Gravitational Physics*, ed. P. F. Michelson (World Scientific, Singapore, 1988); J. W. Armstrong, P. L. Bender, R. W. P. Drever, R. W. Hellings,

R. T. Stebbins, and P.R. Saulson, in *Astrophysics from the Moon, A NASA Workshop, Annapolis, MD 1990*, eds. M. J. Mumma and H. J. Smith (AIP, New York, 1990).

213. B. F. Schutz, *Nature* **323** (1986) 310.

214. C. Cutler, T. A. Apostolatos, L. Bildsten, L. S. Finn, E. E. Flanagan, D. Kennefick, D. M. Markovic, A. Ori, E. Poisson, G. J. Sussman, and K. S. Thorne, *Phys. Rev. Lett.* **70** (1993) 2984.

215. The "no hair conjecture" was proposed by J. A. Wheeler. For a history of its proof, see W. Israel, in *300 Years of Gravitation*, eds. S. W. Hawking and W. Israel (Cambridge University Press, Cambridge, 1987).

216. S. L. Detweiler, *Ap. J.* **239** (1980) 292; E. W. Leaver, *Proc. Roy. Soc. London* **A402** (1986) 285.

217. R. F. Stark and T. Piran, in *Proceedings of the Fourth Marcel Grossman Meeting*, ed. R. Ruffini (Elsevier, New York, 1986).

218. R. L. Garwin, *Physics Today*, December 1974, 9; J. Weber, *Physics Today*, December 1974, 11; R. L. Garwin, *Physics Today*, November 1975, 13; J. Weber, *Physics Today*, November 1975, 13.

219. P. Kafka and L. Schnupp, *Astron. Astrophys.* **70** (1978) 97.

220. E. Amaldi and G. Pizzella, in *Relativity, Quanta, and Cosmology in the Development of the Scientific Thought of Albert Einstein*, vol. 1, eds. M. Pantaleo and F. De Finis (Johnson Reprint Corp., New York, 1979).

221. S. P. Boughn, W. M. Fairbank, R. P. Giffard, J. N. Hollenhorst, E. R. Mapoles, M. S. McAshan, P. F. Michelson, H. J. Paik, and R. C. Taber, *Ap. J.* **261** (1982) L19.

222. W. O. Hamilton, private communication.

223. P. Astone, *et al.*, Proceedings of the Xth Italian General Relativity Conference, Bardonecchia, September 1–4, 1992, unpublished.

224. M. Aglietta, A. Castellina, W. Fulgione, G. Trinchero, S. Vernetto, P. Astone, G. Badino, G. Bologna, M. Bassan, E. Coccia, I. Modena, P. Bonifazi, M.G. Castellano, M. Visco, C. Castagnoli, P. Galeotti, O. Saavedra, C. Cosmelli, S. Frasca, G. V. Pallottino, G. Pizzella, P. Rapagnani, F. Ricci, E. Majorana, D. Gretz, J. Weber, and G. Wilmot, *Nuovo Cimento* **106B** (1991) 1257.

225. B. F. Schutz, talk at 13th International Conference on General Relativity and Gravitation, Huerta Grande, Cordoba, Argentina, 1992.

226. G. Pizzella, in *Developments in General Relativity, Astrophysics and Quantum Theory: A Jubilee Volume in Honour of Nathan Rosen*, eds. F. Cooperstock, L. P. Horwitz, and J. Rosen, *Annals of the Israel Physical Society* **9** (1990) 195.

227. J. Weber, *Found. Phys.* **14** (1984) 1185; J. Weber, in *Gravitational Radiation and Relativity*, Proceedings of the Sir Arthur Eddington Centenary Symposium, Nagpur, India, 1984, eds. J. Weber and T. M. Karade (World Scientific, Singapore, 1986); J. Weber, in *Recent Advances in General Relativity*, eds. A. Janis and J. Porter (Birkhauser, Boston, 1991).

228. L. P. Grishchuk, *Phys. Rev.* **D45** (1992) 2601.

229. M. Hereld, Ph. D. thesis, California Institute of Technology, 1984, unpublished.

230. S. Smith, Ph. D. thesis, California Institute of Technology, 1988, unpublished.

231. M. Zucker, Ph. D. thesis, California Institute of Technology, 1989, unpublished.

232. T. M. Niebauer, A. Rüdiger, R. Schilling, L. Schnupp, W. Winkler, and K. Danzmann, *Phys. Rev.* **D47** (1993) 3106.

233. D. Nicholson, C. A. Dickson, W. J. Watkins, B. F. Schutz, J. Shuttleworth, G. S. Jones, D. I. Robertson, N. L. Mackenzie, K. A. Strain, B. J. Meers, G. P. Newton, H. Ward, C. A. Cantley, N. A. Robertson, J. Hough, K. Danzmann, T. M. Niebauer, A. Rüdiger, R. Schilling, L. Schnupp, and W. Winkler, paper in preparation, 1993.

234. J. Kristian, C. R. Pennypacker, J. Middleditch, M.A. Hamuy, J. N. Imamura, W. E. Kunkel, R. Lucino, D. E. Morris, R. A. Muller, S. Perlmutter, S. J. Rawlings, T. P. Sasseen, I. K. Shelton, T. Y. Steiman-Cameron, and I. R. Tuohy, *Lett. Nature* **338** (1989) 234.

235. J. Kristian, *Lett. Nature* **349** (1991) 747.

236. R. W. Hellings, J. D. Anderson, P. S. Callahan, and A. T. Moffet, *Phys. Rev.* **D23** (1981) 844.

237. R. W. Hellings, in *The Detection of Gravitational Waves*, ed. D. Blair (Cambridge University Press, Cambridge, 1991).

238. S. P. Boughn and J. R. Kuhn, *Ap. J.* **286** (1984) 387.

239. R. W. Hellings and G. S. Downs, *Ap. J. Lett.* **265** (1983) 139; R. W. Romani and J. H. Taylor, *Ap. J. Lett.* **265** (1983) 135.

240. M. M. Davis, J. H. Taylor, J. M. Weisberg and D. C. Backer, *Nature* **315** (1985) 547.

241. L. Krauss and M. White, *Phys. Rev. Lett.* **69** (1992) 869; R. L. Davis, H. M. Hodges, G. F. Smoot, P. J. Steinhardt, and M. S. Turner, *Phys. Rev. Lett.* **69** (1992) 1856.

242. E. Morrison, J. Hough, B. J. Meers, G. P. Newton, D. I. Robertson, K. A. Strain, P. J. Veitch, and H. Ward, in *The Sixth Marcel Grossman Meeting*, eds. H. Sato and T. Nakamura (World Scientific, Singapore, 1992).

243. R. Weiss, *Quarterly Progress Report, MIT Research Lab of Electronics* **105** (1972) 54–76.

244. B. P. Abbott *et al.* (The LIGO Scientific Collaboration and the Virgo Collaboration), *Phys. Rev. Lett.* **116** (2016) 061102.

245. B. P. Abbott *et al.* (The LIGO Scientific Collaboration and the Virgo Collaboration), *Phys. Rev. Lett.* **116** (2016) 131103.

246. B. P. Abbott *et al.* (The LIGO Scientific Collaboration and the Virgo Collaboration), *Phys. Rev. Lett.* **116** (2016) 241103.

247. M. Maggiore, *Gravitational Waves: Volume 1: Theory and Experiments* (Oxford University Press, UK, 2007).

248. J. D. E. Creighton and W. G. Anderson, *Gravitational-Wave Physics and Astronomy: An Introduction to Theory, Experiment and Data Analysis* (Wiley-VCH, Germany, 2011).

249. D. G. Blair, E. J. Howell, L. Ju and C. Zhao (eds.), *Advanced Gravitational Wave Detectors* (Cambridge University Press, UK, 2012).

250. M. Bassan (ed.), *Advanced Interferometers and the Search for Gravitational Waves: Lectures from the First VESF School on Advanced Detectors for Gravitational Waves* (Springer, Switzerland, 2014).

251. D. Reitze and P. Saulson (eds.), *Advanced Gravitational Wave Interferometric Detectors*, Volumes 1 and 2 (World Scientific, Singapore, 2017).

252. *Babylonian Talmud: Tractate Shabbath*, Folio 31a.

253. P. G. Roll, R. Krotkov and R. H. Dicke, *Ann. Phys. (N.Y.)* **26** (1964) 442–517.

254. T. T. Fricke, N. D. Smith-Lefebvre, R. Abbott, R. Adhikari, K. L. Dooley, M. Evans, P. Fritschel, V. V. Frolov, K. Kawabe, J. S. Kissel, B. J. J. Slagmolen and S. J. Waldman, *Class. Quant. Grav.* **29** (2012) 065005.

255. D. Kennefick, *Traveling at the Speed of Thought: Einstein and the Quest for Gravitational Waves* (Princeton University Press, USA, 2007).
256. P. R. Saulson, *Gen. Relativ. Gravit.* **43** (2011) 3289–99.
257. M. Bartusiak, *Einstein's Unfinished Symphony: Listening to the Sounds of Space-time* (Joseph Henry Press, Washington, DC, 2000).
258. J. Levin, *Black Hole Blues and Other Songs from Outer Space* (Alfred A. Knopf, New York, 2016).
259. H. M. Collins, *Gravity's Shadow: The Search for Gravitational Waves* (University of Chicago Press, USA, 2004).
260. H. M. Collins, *Gravity's Ghost and Big Dog* (University of Chicago Press, USA, 2013).
261. H. M. Collins, *Gravity's Kiss* (University of Chicago Press, USA), in press.
262. P. Linsay, P. Saulson and R. Weiss, with contributions by S. Whitcomb, *A Study of a Long Baseline Gravitational Wave Antenna System* (Massachusetts Institute of Technology, USA, 1983).
263. P. R. Saulson, *Am. J. Phys.* **65** (1997) 501–505.
264. D. Garfinkle, *Am. J. Phys.* **74** (2006) 196–199.
265. P. Kafka, in *ESA, Space Science and Fundamental Physics, Proc. Summer School Alpach 1987* (1988), pp. 121-130.
266. F. Pretorius, *Phys. Rev. Lett.* **95** (2005) 121101.
267. M. Campanelli, C. O. Lousto, P. Marronetti and Y. Zlochower, *Phys. Rev. Lett.* **96** (2006) 111101.
268. J. G. Baker, J. Centrella, D.-I. Choi, M. Koppitz and J. van Meter, *Phys. Rev. Lett.* **96** (2006) 111102.
269. J. Abadie *et al.* (The LIGO Scientific Collaboration and the Virgo Collaboration), *Class. Quant. Grav.* **27** (2010) 173001.
270. B. P. Abbott (The LIGO Scientific Collaboration and the Virgo Collaboration), submitted to *Astrophys. J. Lett.* (2016).
271. C. D. Ott, *Class. Quant. Grav.* **26** (2009) 063001.
272. A. Buonanno, Y. Chen and N. Mavalvala, *Phys. Rev. D* **67** (2003) 122005.
273. K. L. Dooley, J. R. Leong, T. Adams, C. Affeldt, A. Bisht, C. Bogan, J. Degallaix, C. Gräf, S. Hild, J. Hough, A. Khalaidovski, N. Lastzka, J. Lough, D. Macleod, L. Nuttall, M. Prijatelj, R. Schnabel, E. Schreiber, J., Slutsky, B. Sorazu, K. A. Strain, H. Vahlbruch, M. Wąs, B. Willke, H. Wittel, K. Danzmann and H. Grote, *Class. Quant. Grav.* **33** (2016) 075009.
274. J. Abadie (The LIGO Scientific Collaboration), *Nature Physics* **7** (2011) 962–965.
275. T. T. Lyons, M. W. Regehr and F. J. Raab, *Applied Optics* **39** (2000) 6761–6770.
276. B. P. Abbott *et al.* (The LIGO Scientific Collaboration), *Rep. Prog. Phys.* **72** (2009) 076901.
277. T. Accadia *et al.* (The Virgo Collaboration), *J. Inst.* **7** (2012) P03012.
278. J. Aasi *et al.* (The LIGO Scientific Collaboration), *Class. Quant. Grav.* **32** (2015) 074001.
279. F. Acernese *et al.* (The Virgo Collaboration), *Class. Quant. Grav.* **32** (2015) 024001.
280. G. I. González and P. R. Saulson, *Phys. Lett. A* **201** (1995) 12–18.
281. K. Numata, M. Ando, K. Yamamoto, S. Otsuka and K. Tsubono, *Phys. Rev. Lett.* **91** (2003) 260602.

282. E. D. Black, A. Villar, K. Barbary, A. Bushmaker, J. Heefner, S. Kawamura, F. Kawazoe, L. Matone, S. Meidt, S. R. Rao, K. Schulz, M. Zhang and K. G. Libbrecht, *Phys. Lett. A* **328** (2004) 1–5.

283. A. Gillespie and F. Raab, *Phys. Rev. D* **52** (1995) 577–585.

284. Y. Levin, *Phys. Rev. D* **57** (1998) 659–663.

285. G. M. Harry, A. M. Gretarsson, P. R. Saulson, S. E. Kittelberger, S. D. Penn, W. J. Startin, S. Rowan, M. M. Fejer, D. R. M. Crooks, G. Cagnoli, J. Hough and N. Nakagawa, *Class. Quant. Grav.* **19** (2002) 897–917.

286. G. M. Harry, H. Armandula, E. Black, D. R. M. Crooks, G. Cagnoli, J. Hough, P. Murray, S. Reid, S. Rowan, P. Sneddon, M. M. Fejer, R. Route and S. D. Penn, *Appl. Opt.* **45** (2006) 1569–1574.

287. D. H. Gwo, *Proc. SPIE* **136** (1998) 136–142.

288. A. Ageev, B. Cabrera Palmer, A. De Felice, S. D. Penn and P. R. Saulson, *Class. Quant. Grav.* **21** (2004) 3887–3892.

289. G. Harry, T. P. Bodiya and R. DeSalvo (eds.), *Optical Coatings and Thermal Noise in Precision Measurement* (Cambridge University Press, UK, 2012).

290. S. Braccini, C. Casciano, F. Cordero, F. Corvace, M. De Sanctis, R. Franco, F. Frasconi, E. Majorana, G. Paparo, R. Passaquieti, P. Rapagnani, F. Ricci, D. Righetti, A. Solinas and R. Valentini, *Meas. Sci. Technol.* **11** (2000) 467–476.

291. J. Giaime, P. Saha, D. Shoemaker and L. Sievers, *Rev. Sci. Instrum.* **67** (1996) 208–214.

292. F. Matichard, B. Lantz, R. Mittleman, K. Mason, J. Kissel, B. Abbott, S. Biscans, J. McIver, R. Abbott, S. Abbott, E. Allwine, S. Barnum, J. Birch, C. Celerier, D. Clark, D. Coyne, D. DeBra, R. DeRosa, M. Evans, S. Foley, P. Fritschel, J. A. Giaime, C. Gray, G. Grabeel, J. Hanson, C. Hardham, M. Hillard, W. Hua, C. Kucharczyk, M. Landry, A. Le Roux, V. Lhuillier, D. Macleod, M. Macinnis, R. Mitchell, B. O'Reilly, D. Ottaway, H. Paris, A. Pele, M. Puma, H. Radkins, C. Ramet, M. Robinson, L. Ruet, P. Sarin, D. Shoemaker, A. Stein, J. Thomas, M. Vargas, K. Venkateswara, J. Warner and S. Wen, *Class. Quant. Grav.* **32** (2015) 185003.

293. M. G. Beker, J. F. J. van den Brand, E. Hennes and D. S. Rabeling, *Journal of Physics: Conference Series* **363** (2012) 012004.

294. M. G. Beker, G. Cella, R. DeSalvo, M. Doets, H. Grote, J. Harms, E. Hennes, V. Mandic, D. S. Rabeling, J. F. J. van den Brand and C. M. van Leeuwen, *Gen. Relativ. Gravit.* **43** (2011) 623–656.

295. B. Abbott *et al.* (The LIGO Scientific Collaboration and the ALLEGRO Collaboration), *Phys. Rev. D* **76** (2007) 022001.

296. E. Hirose, T. Sekiguchi, R. Kumar and R. Takahashi (for the KAGRA collaboration), *Class. Quant. Grav.* **31** (2014) 224004.

297. M. Punturo *et al.* (The Einstein Telescope Collaboration), *Class. Quant. Grav.* **27** (2010) 194002.

298. S. Dwyer, D. Sigg, S. W. Ballmer, L. Barsotti, N. Mavalvala and M. Evans, *Phys. Rev. D* **91** (2015) 082001.

299. B. P. Abbott *et al.* (The LIGO Scientific Collaboration and the Virgo Collaboration), *Phys. Rev. D* **93**(2016) 122003.

300. B. P. Abbott *et al.* (The LIGO Scientific Collaboration and the Virgo Collaboration), *Phys. Rev. D* **93** (2016) 122004.

301. B. P. Abbott *et al.* (The LIGO Scientific Collaboration and the Virgo Collaboration), Comprehensive all-sky search for periodic gravitational waves in the sixth science run LIGO data (2016), arXiv:1605.03233.

302. J. Aasi *et al.* (The LIGO Scientific Collaboration and the Virgo Collaboration), *Phys. Rev. Lett.* **113** (2014) 231101.

303. B. Allen, *Phys. Rev. D* **71** (2005) 062001.

304. J. Aasi *et al.* (The LIGO Scientific Collaboration and the Virgo Collaboration), *Class. Quant. Grav.* **32** (2015) 115012.

305. B. P. Abbott *et al.* (The LIGO Scientific Collaboration and the Virgo Collaboration), Characterization of transient noise in Advanced LIGO relevant to gravitational wave signal GW150914 (2016), arXiv:1602.03844.

306. B. P. Abbott *et al.* (The LIGO Scientific Collaboration and the Virgo Collaboration and the ASKAP Collaboration, the BOOTES Collaboration, the Dark Energy Survey and the Dark Energy Camera GW-EM Collaborations, the *Fermi* GBM Collaboration, the *Fermi* LAT Collaboration, GRAWITA, the *INTEGRAL* Collaboration, the iPTF Collaboration, the InterPlanetary Network, the J-GEM Collaboration, the La Silla – QUEST Survey, the Liverpool Telescope Collaboration, the LOFAR Collaboration, the MASTER Collaboration, the MAXI Collaboration, the PESSTO Collaboration, the Pi of the Sky Collaboration, the SkyMapper Collaboration, the *Swift* Collaboration, the TAROT, Zadko, Algerian National Observatory and C2PU Collaboration, the TOROS Collaboration, and the VISTA Collaboration), Localization and broadband follow-up of the gravitational-wave transient GW150914, *Astrophys. J. Lett.* (2016), in press.

307. B. P. Abbott *et al.* (The LIGO Scientific Collaboration and the Virgo Collaboration), *Living Rev. Relativity* **19** (2016) 1.

308. J. Abadie *et al.* (The LIGO Scientific Collaboration and the Virgo Collaboration and M. S. Briggs, V. Connaughton, K. C. Hurley, P. A. Jenke, A. von Kienlin, A. Rau and X.-L. Zhang), *Astrophys. J.* **760** (2012) 12.

309. B. P. Abbott *et al.* (The LIGO Scientific Collaboration and the Virgo Collaboration), *Phys. Rev. Lett.* **116** (2016) 241102.

310. B. P. Abbott *et al.* (The LIGO Scientific Collaboration and the Virgo Collaboration), *Phys. Rev. Lett.* **116** (2016) 221101.

311. B. P. Abbott *et al.* (The LIGO Scientific Collaboration and the Virgo Collaboration), *Astrophys. J. Lett.* **818** (2016) L22.

312. M. Armano *et al.* (The LPF Collaboration), *Phys. Rev. Lett.* **116** (2016) 231101.

313. G. Hobbs *et al.* (The IPTA Collaboration), *Class. Quant. Grav.* **27** (2010) 084013.

314. P. A. R. Ade *et al.* (The BICEP2/Keck and Planck Collaborations), *Phys. Rev. Lett.* **114** (2015) 101301.

Printed in the United States
By Bookmasters